ADVANCES IN
DIGITAL IMAGE
PROCESSING
Theory, Application, Implementation

THE IBM RESEARCH SYMPOSIA SERIES

ADVANCES IN DIGITAL IMAGE PROCESSING

Theory, Application, Implementation

Edited by
Peter Stucki

IBM Zurich Research Laboratory
Zurich, Switzerland

PLENUM PRESS · NEW YORK AND LONDON

Library of Congress Cataloging in Publication Data

International Symposium on Advances in Digital Image Processing, Bad Neuenahr,
  Ger., 1978.
  Advances in digital image processing.

  (The IBM research symposia series)
  "Proceedings of the International Symposium on Advances in Digital Image Pro-
cessing, held at Bad Neuenahr, Federal Republic of Germany, September 26—28,
1978."
  Includes index.
  1. Image processing—Congresses. I. Stucki, Peter. II. Title. III. Series: International
Business Machines Corporation. IBM research symposia series.
  TA1632.I59 1978   621.3815'42   79-21443
  ISBN 0-306-40314-5

Proceedings of the International Symposium on
Advances in Digital Image Processing, held at Bad Neuenahr,
Federal Republic of Germany, September 26—28, 1978.

© 1979 Plenum Press, New York
A Division of Plenum Publishing Corporation
227 West 17th Street, New York, N.Y. 10011

Printed in the United States of America

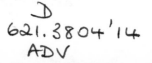

PREFACE

Digital image processing, originally established to analyze and improve lunar images, is rapidly growing into a wealth of new applications, due to the enormous technical progress made in computer engineering. At present, the most important fields of growth appear to emerge in the areas of medical image processing (i.e. tomography, thermography), earth resource inventory (i.e. land usage, minerals), office automation (i.e. document storage, retrieval and reproduction) and industrial production (i.e. computer vision for mechanical robots). Currently, emphasis is being shifted from signal-processing research and design-innovation activities towards cost-efficient system implementations for interactive digital image processing. For the years ahead, trends in computer engineering indicate still further advances in Large Scale Integration (LSI) and Input/Output (I/O) technologies allowing the implementation of powerful parallel and/or distributed processor architectures for real-time processing of high-resolution achromatic and color images.

In view of the many new developments in the field of digital image processing and recognizing the importance of discussing these developments amongst key scientists that might make use of them, IBM Germany sponsored an international symposium on 'Advances in Digital Image Processing', held at Bad Neuenahr, Federal Republic of Germany, September 26 - 28, 1978.

The interest shown in this symposium encouraged the publication of the papers presented in this volume of the IBM Research Symposium Series. It is felt that these proceedings will serve as a useful reference for all those involved in digital image processing, either on a practical or a theroretical level.

Zurich and Stuttgart, April 1979   Peter Stucki
                                         IBM Zurich Research Laboratory
                                         Symposium Chairman

                                         Paul Schweitzer
                                         IBM Germany, Stuttgart
                                         Symposium Manager

Table of Contents

APPLICATION

IMPLEMENTATION

A LOW-COST IMAGE PROCESSING FACILITY EMPLOYING A NEW HARDWARE REALIZATION OF HIGH-SPEED SIGNAL PROCESSORS

GENERAL

EVOLUTION IN IMAGE SCIENCE

E. Klein and H. J. Metz,
Agfa-Gevaert AG
Leverkusen
Federal Republic of Germany

1. INTRODUCTION

There exists a great number of principles, methods and technologies suitable for recording, storage and reproduction of optical information in form of continuous-tone images.

Silver halide photography, halftone printing and television represent the most important classical systems. Digital image processing comprises some more recent methods of handling, transformation and enhancement of digitized images with the aid of computing systems. Image science looks at the problems encountered with the handling of optical information from a general point of view.

The first part of the paper gives a short review of the silver halide system. The second part discusses some general features of imaging systems from the point of view of image science.

2. THE SILVER HALIDE SYSTEM

2.1. General features

Silver halide microcrystals in the size range of 0,1 to 1 μm constitute the light sensitive elements of the

silver halide system. They are statistically distributed within a gelatin layer. According to different
methods of production the crystals may show a considerable variation in the distribution of grain size and
shape. Figure 1 shows, from the left to the right, a
conventional heterodisperse emulsion, a monosized cubic
emulsion and a monosized emulsion of octahedral habit.

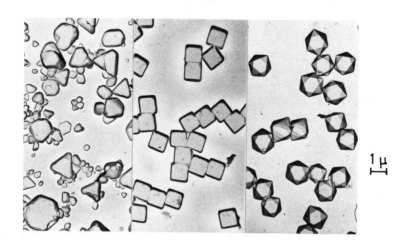

Fig. 1   Different types of photographic silver halide
         emulsions (electron micrograph of carbon
         replica).

Upon exposure to light a silver halide crystal may
produce one or more "latent image"-centers which in a
subsequent development process efficiently catalyze
the transformation of the whole grain into a cluster
of metallic silver. For optimum conditions the absorption of three or perhaps just two quanta of light
seems to be sufficient for the formation of a develop-
able latent image center. In general, however, a considerable higher number of quanta has to be absorbed
in order to reach the state of developability. Number
and position of latent image centers may be rendered
visible by special decoration techniques. The result
of such a procedure is shown in Fig. 2.

Latent image formation may be regarded as a stochastic
process. Theoretically, any individual crystal might

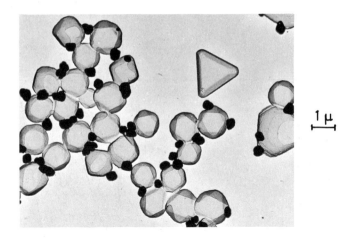

Fig. 2 Decoration of latent image centers by con-
         trolled development.

thought to be characterized by its own response function which gives the probability of activation (by
formation of a latent image) as a function of, e.g.,
the mean number of quanta absorbed. Experimentally a
response function cannot be determined for a single
crystal but for a class of grains of equal size and
shape, e.g., for a monodisperse emulsion. In Fig. 3 such
a response function (dotted curve), measured for a
monodisperse emulsion, is compared to a set of theoretical curves derived for the assumption of a fixed
activation threshold r.

The macroscopic response curve of a given emulsion
layer - density vs. exposure - might be obtained by
weighted superposition of the primary response curves
for the different grain classes, taking into account
the size dependent transformation of absorbed quanta
axis into exposure axis as well as the influence of the
depth dependence of exposure within a thick photographic layer.

In general, the response function is not just a function of the exposure E (irradiance I times exposure
time t) but is explicitly dependent on the exposure
conditions, e.g., on the level of irradiance I. Figure 4
shows the dependence on the irradiance of the exposure
necessary to produce a given response, for a unsensitized monodisperse emulsion. The existence of "reci-

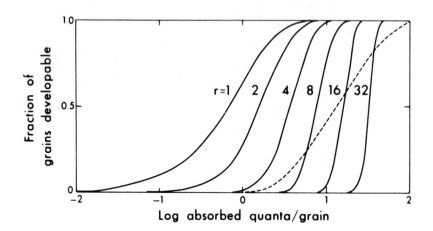

Fig. 3   Microscopic response function (theoretical and experimental) for monodisperse emulsion (ref. [1]).

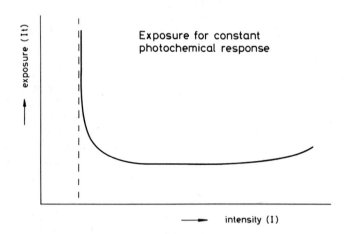

Fig. 4   Reciprocity failure for a unsensitized monodisperse emulsion.

procity failure" leads to the corollary that a latent
image center cannot be the result of absorption of just
one photon nor of <u>independent</u> reactions of more than
one photon. Instead a combined action of at least two
photons has to be assumed. Accordingly, high-speed
emulsions can only be realized with large silver halide
crystals providing a high absorption cross section.
With a one-quantum process high speed could be achieved
by the employment of a sufficient high concentration
of arbitrarily small elements, e.g., molecules. The only
problem left would be to find a compromise between
practical speed (determined here by the factor of amplification) and sufficient storage stability of the
unexposed material on the shelf.

Amplification means attaching to a primary latent image
a large optical cross section in order to render it
"visible". High gain amplification is easily achieved
in the silver halide system by transformation of the
activated grains into clusters of metallic silver.
Figure 5 gives an example for a further amplification of
a faint silver-cluster image by a catalytic reaction,
which produces light-scattering oxygen microbubbles by

Fig. 5 Amplification of a faint silver image by
catalytic desintegration of H_2O_2.

desintegration of H_2O_2 at the site of the developed silver.

The long wavelength edge of the absorption of silver halide is situated somewhere in the blue spectral region. Radiation of longer wavelength is not absorbed and hence not recorded by the untreated silver halide crystal. Adsorption of certain "sensitizing" dyes to the silver halide surface may extend the range of sensitivity to the green, red or to both these parts of the spectrum (Fig. 6). In the ideal case every light quantum absorbed by the dye leads to a conduction electron in the silver halide crystal.

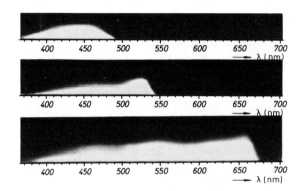

Fig. 6   Spectral sensitization of silver halide
          Response of unsensitized, ortho-sensitized
          (green) and pan-sensitized (green + red)
          emulsion.

The silver halide system owes its still outstanding position as a primary image-recording material to a rather unique combination of different advantageous properties: a high quantum sensitivity of the AgHal grain, the existence of a low-intensity reciprocity failure, a built-in high gain amplification and the possibility of spectral sensitization.

2.2. Colour Photography

Optical sensitization of silver halide crystals enables
the recording of different spectral parts of a radiation image at different sites of a suitably designed
layer system. This is the basic requirement for colour
photography.

There are two fundamentally different systems possible.
The additive system employs colour screens containing
equal areas of the principal colours blue, green and
red. Colour reproduction is achieved by means of a pan-
sensitized black and white layer combined with the additive screen for recording and reproduction (slide
projection).

However, more advantageous than the additive system is
the subtractive system which records the blue, green
and red part of the image in different layers of a
multilayer arrangement thus producing the correspondent
complementary dyes yellow, magenta and cyan, respectively. Similar to the black-and-white technique, in a
first step a negative image of complementary colour is
produced, which in a second step may be copied onto a
similar material yielding the final positive image.

Chemically the dyes within the different layers may be
produced according to the development of the silver
halide crystals by making use of the oxidized developer
molecule. A scheme of this oxidative dye-coupling process is shown in Fig. 7. The exposed silver halide
crystal is reduced by a substituted p-phenylen-diamine
such producing clusters of metallic silver and oxidized
developer molecules. The latter are able to react with
a corresponding colourless coupler molecule within the
layer forming the corresponding dyes of the substrac-
tive system.

Based on the subtractive system, considerable improvement in colour reproduction has been achieved during
the last decades e.g., by introducing of an automatic
masking system using coloured couplers in order to
compensate for the unwanted secondary absorption of
actual dyes. Definition and colour reproduction may be
further controlled by deliberately making use of inter-
layer and interimage effects.

Certainly, the most striking development in the field
of colour photography is the integral instant colour

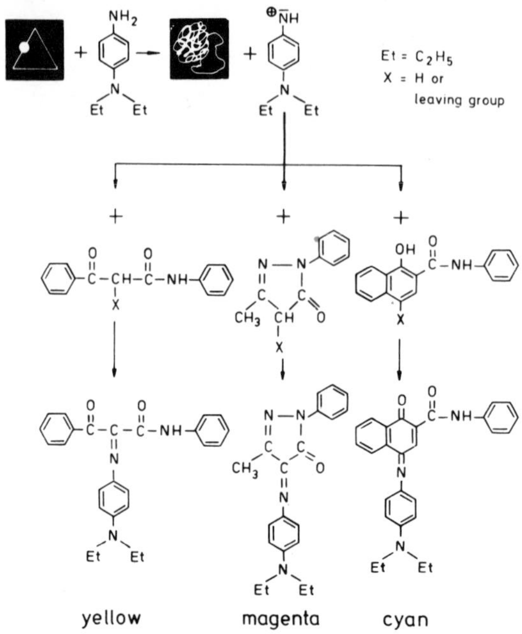

<u>Fig. 7</u>   Scheme of colour development
             (Oxidative dye-coupling process).

system which yields the final image within some minutes
after exposure. In this case also the dark-room and the
whole processing chemistry have been incorporated into
the material. The basic working principle relies upon a
development-controlled dye diffusion process which
leads to an imagewise transport of dyes to an image-
recording layer that is optically separated from the
image-recording part of the set.

It is interesting that two principally different reali-
zations of the system have been found. The first one,
given in Fig. 8, was introduced by Polaroid in 1972.
It involves negative working emulsions. Exposure and
view of the final image are from the same side of the
set, therefore, a special optical arrangement in the
camera is necessary in order to expose a mirrored image.

In the second system (Kodak 1976) a conventional image
is recorded at one side and the view onto the final
image is from the other. It is working on special reversal emulsions.

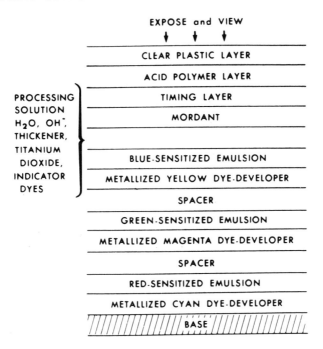

EXPOSE and VIEW

CLEAR PLASTIC LAYER

ACID POLYMER LAYER

PROCESSING
SOLUTION
H_2O, OH^-,
THICKENER,
TITANIUM
DIOXIDE,
INDICATOR
DYES

TIMING LAYER

MORDANT

BLUE-SENSITIZED EMULSION

METALLIZED YELLOW DYE-DEVELOPER

SPACER

GREEN-SENSITIZED EMULSION

METALLIZED MAGENTA DYE-DEVELOPER

SPACER

RED-SENSITIZED EMULSION

METALLIZED CYAN DYE-DEVELOPER

BASE

Fig. 8   Schematic arrangement of layers in the integral
         instant film unit (Polaroid, ref. [1]).

2.3. Equidensitometry

In the course of its hundred years history, the silver
halide system has been adapted to a large variety of
special applications. One of these special adaptations
was developed by Agfa-Gevaert only some years ago.
The material (Agfacontour-film) enables the quantitative evaluation of images by means of equidensitometry,
which is made possible due to a very unique response
characteristic given in Fig. 9.

This response curve is achieved by a somewhat involved
interaction between a silver bromide and a silver
chloride emulsion which are present in the same layer.

The narrow, V-shaped response curve attaches a significant response only to a narrow exposure region. When a
copy of a given spatial density distribution is made
onto this material, with a given exposure a certain
small density region is selected from the original.
By a second copying process on the same kind of material

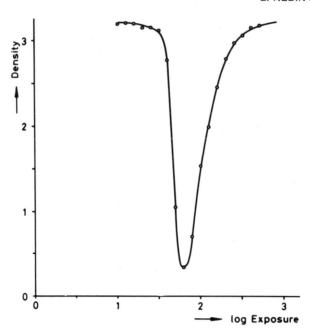

Fig. 9   Response curve of Agfacontour-film.

this density region may be marked as a pair of narrow lines (second-order equidensity lines).

A set of different exposures, made e.g., at constant log-exposure steps, and subsequent collection on a single copy yields a stratification of the primary density distribution e.g., into a family of second-order equidensity lines. This enables a quantitative evaluation of the image as illustrated in Fig. 10 which shows a conventional astronomic photograph of a comet in the upper part of the figure and the resolution into second-order equidensity lines in the lower part.

A stratification of the density profile of an image may also be obtained by a family of first-order exposures. In order to discriminate between the different density regions these regions may be arbitrarily connected to different colours. An example of this technique is given in Fig. 11, which resolves the density profile within a Laue-diagram.

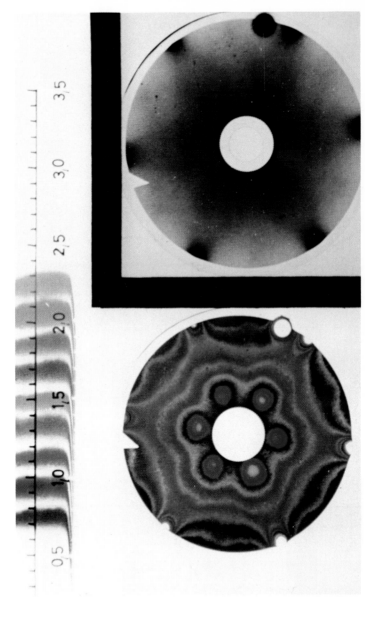

Fig. 11. Transformation of density into colour (Laue-diagram of untwinned mono-crystal of Brazil quartz).

Special screen effects are obtained when e.g., a contact
screen is placed between the original and the equiden-
sity-film. According to different exposures different
density regions in the screen are indicated. Hence,
different geometrical structures are attached to different density regions in the original. A simple application of this method is the direct recording of the irradiance distribution on an area illuminated by a reflector (Fig. 12).

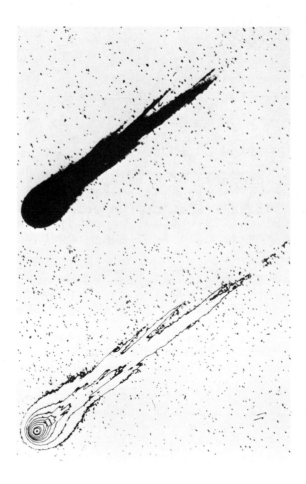

Fig. 10   Transformation of a density profile (comet)
          into a family of second-order equidensity
          lines.

Screen Equidensitometry

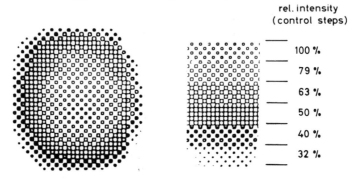

Direct Registration and Quantization
of Light Distribution (Reflector)

rel. intensity
(control steps)

100 %

79 %

63 %

50 %

40 %

32 %

Fig. 12   Transformation of density (or irradiance)
into structure.

3. PROBLEMS OF IMAGE SCIENCE

Generally, any property of a material or system that
can be influenced by irradiation with light may be used
for recording of an image. Any material or device that
can be modified in its optical properties according to
a given image may be used for image reproduction or
display.

Figure 13 provides a general scheme for the handling of
optical information. There are two different ways of
recording an optical image: In the first the exposure
leads to a direct modification of a suitable substrate
which perhaps in a second step may be stabilized, am-
plified, transformed to a visible modification of the
same or another substrate to yield the reproduced
image. The second, indirect way involves the "measurement" of the image e.g., by scanning with a suitable
device which yields a signal string that may be trans-
ferred, stored, processed before the recorded image is
reproduced with the aid of a suitable display unit.

Figure 14 gives a classification of imaging systems.
Here it is assumed that an imaging system may be regarded as consisting of a sufficient large number of independent radiation detectors or image elements respectively. A classification into four basic types is

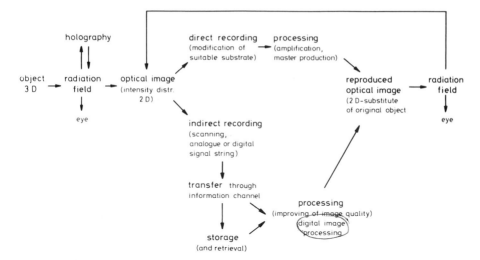

<u>Fig. 13</u> Handling of optical information.

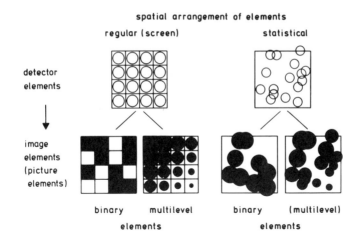

<u>Fig. 14</u> Classification of imaging systems.

achieved according to nature (on-off, multilevel) and
to spatial arrangement (regular, at random) of the
elements.

Generally, recording or reproduction of an optical
image with different systems corresponds to different
techniques of encoding of the continuous-tone image
(Fig. 15).

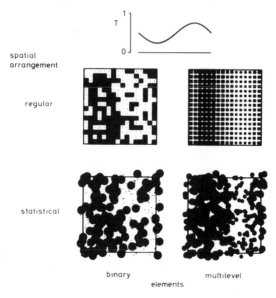

Fig. 15   Performance of model imaging systems.

In the case of stochastic encoding intermediate image tones can only be achieved by a statistical mixture of activated and inactivated elements. The image suffers from "distribution noise". Multilevel encoding (PCM) employs elements of a more involved structure which are able to record or display a continuous-tone value approximately by assuming one of a certain number of possible output values. The image may suffer from insufficient tonal resolution if elements with a low number of levels are employed, but it will not suffer from noise.

Statistical distribution of the elements introduces another source of "distribution noise" (lower part of figure). The relative degradation of the image by this additional noise is tolerable for the stochastic system; it is intolerable for the multilevel system which is totally spoiled.

Still another source of noise becomes important for image recording systems of highest sensitivity. Actually systems do not record a continuous intensity

function, but they record the stochastically encoded
intensity function given as the spatial distribution
of photons absorbed during the exposure interval. As a
consequence, the differences between the different
types of imaging systems given in Fig. 15 essentially
disappear in the theoretical limit of sensitivity where
the photon distribution is exactly recorded.

Figures 16 and 17 are to illustrate the performance of
multilevel system and stochastic system at the reproduction stage (no influence of photon noise) by a pictorial example.

Fig. 16   Reproduction of a continuous-tone picture
           by a 3·3 set of different multilevel systems.

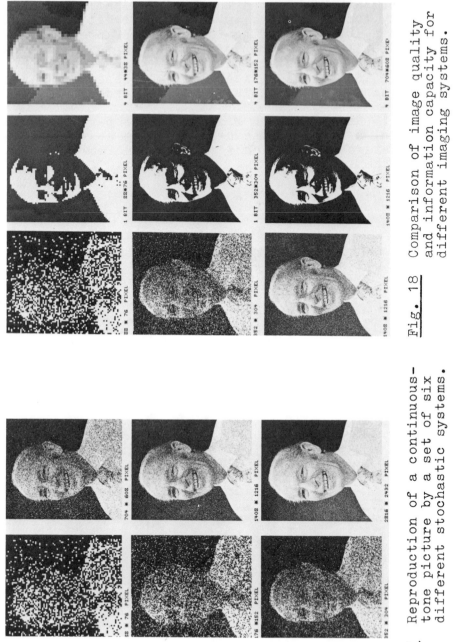

Fig. 18   Comparison of image quality and information capacity for different imaging systems.

Fig. 17   Reproduction of a continuous-tone picture by a set of six different stochastic systems.

Figure 16 shows the result of reproducing a portrait by a set of nine different regular multilevel systems. The number of available levels is 2, 8 and 32 for the first, second and third column of the array. The number of elements is 418 for the first row, it is 16 and 256 times greater for the second and third row.

Acceptable results in image quality are only achieved if both, the number of levels and the number of elements reach at least certain minimum values, lying here somewhere between the 4 pictures in the lower right corner of the array.

Figure 17 shows the same picture reproduced by a set of 6 regular stochastic systems. The number of elements is 6688 for the picture of poorest quality and is multiplied by 4 for every step in the quality scale.

After comparison of Figs. 16 and 17 it is obvious, that technologies used to display or reproduce a recorded image - i.e. printing technologies - generally should be based on the regular multilevel system. The only serious exception from this rule seems to be the silver halide system, which produces a final image by primary recording and successive conversion of the recording elements into image elements.

Information capacity according to the number of different possible images is easily determined for a regular system: $c = n \, ld \, m$, n and m indicating the number of elements and the number of levels, respectively.

Figure 18 shows reproductions made with the regular stochastic system (first column) and with the multilevel system with 2 and 16 levels for the second and third column. The pictures in each row are equal in information capacity which is 6688 bits for the first row and is multiplied by 16 and 256 for the second and third row.

The visual impression obtained with the different system demonstrates that information capacity cannot be accepted as a general measure of image quality if different imaging systems are to be compared.

<u>Conclusion</u>: The four basic types of imaging systems
may be reduced to essentially two different types:
i.e., the regular multilevel system at one side and the
stochastic system at the other.

For the recording stage one may employ either the multilevel system with relatively few large elements of an
involved structure (physical system, e.g., CCD) or the
stochastic system with a considerably greater number of
small on-off detectors. These need not to be regularly
arranged and may be produced by chemical reactions (chemical system, e.g., silver halide).

Neither of these systems, if compared under reasonable
conditions, should be superior to the other for principal reasons. Which one will be the most suitable for
a given task will largely depend on the boundary conditions connected with the problem at hand and with
the state of technical development of the different
systems.

For the reproduction stage, on the other hand, the multilevel system, obviously, has the advantage over the
stochastic system; it is exclusively used for any kind
of halftone-printing where continuous-tone image is
screened up in order to be reproduced by a material
having no inherent gray-scale ability.

4. LITERATURE

[1]   James, T.H., The Theory of the Photographic Pro-
      cess, Fourth Edition, Macmillan, New York, London
      (1977)

[2]   Klein, E., Ber. Bunsenges. phys. Chem. <u>80</u>, 1083
      (1976)

[3]   Metz, H.J., S. Ruchti and K. Seidel, J. Phot. Sci.
      <u>26</u>, 229 (1978)

TRENDS IN DIGITAL IMAGE PROCESSING RESEARCH

T. S. Huang
Purdue University
West Lafayette, Indiana, U. S. A.
and
Universitaet Hannover
Hannover
Federal Republic of Germany

ABSTRACT

We first review briefly the four major areas of image processing : computer graphics, efficient coding, restoration and enhancement, and pattern recognition. Then we discuss some general
trends in digital image processing research.

I. INTRODUCTION

Activities in digital image processing have been increasing
rapidly in the past decade. This is not surprising when one real-
izes that in a broad sense image processing means the processing
of multidimensional signals and that most signals in the real
world are multidimensional. In fact, the one-dimensional signals
we work with are often collapsed versions of multidimensional
signals. For example, speech is often considered as a one-dimensional signal, viz., a function of a single variable (time). However, speech originally exists in space and therefore is a function of 4 variables (3 spatial variables and time).

There are analog (optical, electro-optical) as well as digital techniques for image processing. Because of the inherent advantages in digital techniques (flexibility, accuracy), and because of the rapid progress in computer and related technologies, such as LSI, and VLSI, it is fair to say that except for some very specialized problems, digital techniques are usually preferred.

There are four major areas in digital image processing : computer graphics, efficient coding, restoration and enhancement, and pattern recognition. In this paper, we shall briefly review each of these four areas, and then discuss the recent trends in digital image processing research. Many detailed examples of image processing are given in the other papers in this book.

II. COMPUTER GRAPHICS

Computer graphics deals with the problem of generating and displaying images from computers. A salient feature of computer graphics is that the computer can synthesize objects which do not exist in the real world. For example, one application of computer graphics is the display of perspective views of automobile bodies which have been designed but not built. Mathematical models of the automobile bodies are fed into the computer. Based on these models the computer generates the perspective views.

In displaying perspective views of three-dimensional objects, there are two main problems. The first is hidden surface elimination : to decide which part of the object is occluded and should not be seen. The second is shading : to decide what gray level should be displayed at each point of the screen. The shading depends on not only the object model but also the illumination model. Algorithms for solving both problems have been developed. One aspect of the problem which has not yet received sufficient attention is the generation of realistic textures based on statistical models. This will have application in problems such as flight simulation.

An intriguing problem in computer graphics is the display of three-dimensional objects in space using holographic techniques. This is possible in principle. In fact, computer generation of $360°$ - view holograms has been achieved by Dr. Jaroslavsky, Institute for Automatic Transmission, Academy of Science, U.S.S.R. However, the large resolution requirement of the hologram and the lack of reliable display media dim the prospect of an economical three-dimensional digital holographic display system, at least for the near future.

III. EFFICIENT CODING

In many applications, one needs to transmit or store images in digital form. The number of bits involved is often tremendous. It is desirable and in many cases mandatory to compress or efficiently code the data. For example, each of LANDSAT I, II, and III provides approximately 1.3×10^{13} bits of image data per year. With more spectral bands and higher resolution, LANDSAT D will give approximately 3.7×10^{15} bits per year. All these data have to be stored. Some compression is highly desirable.

In the past, much of the research in image coding was motivated by video phone. More recently, attention has been shifted to teleconferencing and to a lesser extent studio television applications.

Straight digitization of an image requires about 8 bits per picture element (pel). Using in-frame coding techniques (DPCM, transform coding), one can reduce the bit rate to around 1-2 bits per pel which preserves good picture quality. Further reduction is possible if one uses frame-to-frame coding techniques where the correlation between frames is utilized. A way of doing that is to transmit frame differences. Obviously, if there is no motion, the differences will be very small (zero, if there is no noise). But if there is a lot of motion, the differences can be large. An area of active current research is motion-compensated coding. The motion is detected and estimated. Then motion-compensated frame differences are transmitted. A word of caution is in order here. Most coding schemes degrade the image quality. Therefore, the bit rate or compression factor of a coding method is meaningless by itself. One has also to look at the received images to see how much degradation has been incurred.

A completely different class of images is graphics, which includes business letters and documents, engineering drawings, weather maps, etc. These are images of man-made symbols. Most of these images are nominally two-level, i.e., each pel is either black or white - there is no need for intermediate gray levels. The digital storage and transmission of these images have become increasingly important, because of the intensive current interest office automation. Most coding methods of digital graphics are based on runlength coding and its extensions. However, the current trend is in exploring the use of pattern recognition techniques.

IV. RESTORATION AND ENHANCEMENT

Both restoration and enhancement aim at improving image quality. The demarcation between the two is not clear. Roughly speaking, when we talk about restoration, we have some definite degradations (e.g., defocusing) in mind and we want to remove these
degradations to obtain an ideal image (an image which we would have
obtained in the absence of these degradations). Enhancement, on
the other hand, is broader in scope. Here, we want to put the
image in a form that is suitable for our purpose. We do not necessarily want the ideal image. For example, we may want to over-
sharpen the edges, or to use false colors.

Perhaps the most prominent example of successful image restoration is the work done at the Jet Propulsion Laboratory with images
of the moon, Mars, and other planets taken by TV cameras on board
artificial satellites. They have succeeded in compensating for
various image degradations including : random noise, interference,
geometrical distortion, field nonuniformity, contrast loss, and
blurring. Another problem which has spurred much of the research
in image restoration but to which satisfactory solutions are yet
to be found is imaging through the atmosphere. The problem is
to obtain good images of planets, stars, and artificial satellites
by imaging systems based on earth. The dominating degradation in
this case is the blurring due to atmospheric turbulence.

There are two approaches to image restoration : "a priori"
and "a posteriori". In the a priori approach, one tries to invent
novel imaging schemes which would give good images to start with.
In the a posteriori approach, one uses conventional imaging systems
such as a camera, and then attempts to improve the image thus obtained. In the case of imaging through the atmosphere, several
a priori methods have been proposed. These include holographic
techniques and coded-aperture imaging - the latter has achieved
some success. An area of intensive current research is adaptive
optics. Here, the phase disturbance due to the atmosphere is
measured, and controllable deformable optics are used to compensate
for the disturbance in real-time.

In the a posteriori approach, most of the effort has been
concentrated on linear degradations. This is because of the
fact that most image degradations can be modeled adequately by
a cascade of three subsystems: a linear part, a nonlinear no-
memory part, and noise. For example, such a model can be used for
film, where the linear part represents blurring due to chemical
diffusion, the nonlinear no-memory part represents the H-D curve,
and the noise is due to film grains. The term "no-memory" is
used in the following sense. A two-dimensional system is said to

have no memory, if the value of the output at any given point depends only on the value of the corresponding point in the input
but not on the values of any other input points. If the characteristics of the degradations are known, then compensation for
the nonlinear no-memory part is straightforward. The more difficult task is to compensate for or equivalently to find the inverse
of the linear degradation. The performance of image restoration
methods is ultimately limited by the presence of noise.

    If the linear degradation is spatially-invariant, then Fourier
techniques (e.g., Wiener inverse filtering) can be used to do the
inversion. Examples of linear spatially-invariant degradations include defocusing and spherical aberration of the lens, and trans-
lational motion of the camera relative to the object. In the
Fourier techniques, one applies a linear spatially-invariant (LSI)
inverse filter to the degraded image to obtain the restoration.
LSI inverse filters have severe limitations. Since such a filter
multiplies each spatial frequency component of the degraded image
by a complex constant, it cannot alter the signal-to-noise ratio
at any individual spatial frequency. The only thing it can do is
to boost up those frequency components where the signal is larger
than the noise and suppress those frequency components where the
noise is larger than the signal. A more dramatic implication is
that there is no way for such a filter to recover those spatial-
frequency components of the ideal image which were cut off by a
band-limiting degrading system - this fact is sometimes stated as :
"LSI inverse filters cannot achieve superresolution."

    In order to achieve superresolution, one has to use nonlinear
(NL) or linear spatially-varying (LSV) techniques, and to have
additional information about the signal. For noncoherent imaging
systems, one piece of information we always have : the image is
non negative. Another piece of information which is not always
true is that the scene giving rise to the image may consist of a
spatially-limited object lying in a uniform background.

    A number of LSV and NL restoration methods have been developed
which make use of these pieces of information. They fall into two
categories : Frequency-domain and spatial-domain techniques. The
latter have been proven more successful mainly because the information about the signal is stated in the spatial-domain and it
is very awkward to try to translate it into the frequency domain.
The spatial domain techniques have the further advantage that they
can be applied to linear spatially-varying degradations (such as
lens aberrations except spherical aberration, and rotational motion
of the camera relative to the object) as well.

Among LSV techniques, several are based on generalized matrix inverses. These include singular-value decomposition, and projection iterative method. NL techniques include maximum-entropy methods and Bayes estimation. Generally, these methods can achieve much better restoration than Fourier methods. However, one has to pay the price of computational complexity. For example, applying any of these spatial-domain methods to even a small image (128x128 points, say) can take several hours on a mini-computer (e.g., PDP 11/45 with floating-point hardware). An important problem to investigate is then : How to improve the computational efficiency of these methods?

V. PATTERN RECOGNITION

In a narrow sense, pattern recognition means the classification of a given unknown pattern into one of a number of standard classes. This is normally done in two steps. First, features are extracted from the given pattern. These features are represented by numbers. The vector with these features as components is called the feature vector. Second, based on the feature vector a classification is made. Usually, training samples are used to design the classifier.

The oldest area of application of pattern recognition is probably OCR (Optical Character Recognition). More recently, pattern recognition has found successful applications in remote sensing (e.g., land use classification of LANDSAT images), biomedical problems (e.g., classification of white blood cells), and industrial problems (e.g., quality control).

In a broad sense, pattern recognition means scene analysis. We want to build an automatic or man-machine interactive system which would look at a scene and derive a symbolic description of it. The process of transforming raw image data into symbolic representation is a complex one; therefore, we subdivide it into several steps as shown in Fig. 1. We first consider the left side of the block diagram in Fig. 1. After the sensor collects the image data, the preprocessor may either compress it for storage or transmission or it may attempt to put the data into a form more suitable for analysis. Image segmentation may simply involve locating objects in the image or, for complex scenes, determination of characteristically different regions. Each of the objects or regions is categorized by the classifier which may use either classical decision-theoretic methods or the more recently developed syntactic methods. In linguistic terminology, the regions (objects) are primitives, and the classifier finds attributes for these primitives. Finally, the structural analyzer attempts to determine the spatial, spectral, and/or temporal relationships among the

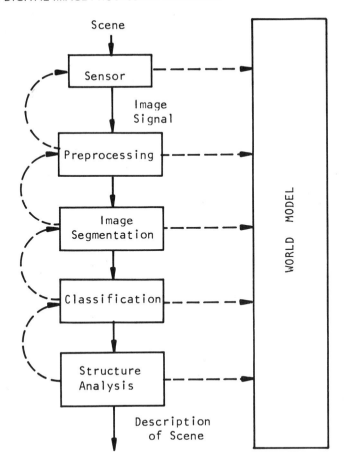

Fig. 1 A scene analysis system.

classified primitives. The output of the "Structure Analysis"
block will be a description (qualitative as well as quantitative)
of the original scene. Notice that the various blocks in the
system are highly interactive. Usually, in analyzing a scene one
has to go back and forth through the system several times.

Past research has indicated that scene analysis can be successful only if we restrict a priori the class of scenes we are
analyzing. This is reflected in the right side of the block
diagram in Fig. 1. A world model is postulated for the class of
scenes at hand. This model is then used to guide each stage of
the analyzing system. The results of each processing stage can
be used in turn to refine the world model.

Before we start to analyze a scene, a world model is constructed which incorporates as much a priori information about the
scene as possible. This could, for example, be in the form of a
relational graph containing unknown parameters. Then the analysis
problem becomes the determination of these unknown parameters. In
this way, the difficult problem of scene analysis is reduced to
the (conceptually) much simpler problems of detection, recognition,
and mensuration.

In the past, scene analysis was done only on extremely simple
images, e.g., the block world. More recently, there is considerable activity in trying to apply a combination of techniques from
pattern recognition and artificial intelligence to the analysis of
complex scenes (such as the interior of a room, an aerial photo of
an airport, etc.).

VI. GENERAL TRENDS

We have seen some of the trees of digital image processing.
Now, let us back off and look at the forest as a whole and to discuss some of the general trends in digital image processing research.

To solve a problem in image processing, we ideally go through
three stages as depicted in Fig. 2. We first come up with a
mathematical model of the image, and in the case where the processed image is to be viewed by humans also a mathematical model of
the human visual process in so far as how it judges the performance
of the processing. Then based on these models, we try to develop
optimum mathematical techniques to do the processing. Finally,
the mathematical techniques are implemented efficiently by either
software or special-purpose hardware.

For example, in restoring an image degraded by lens defocusing,
the mathematical model for the image has two parts : the signal is
characterized by a stochastic process with certain energy spectrum;

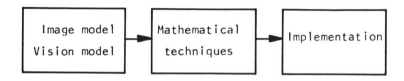

Fig. 2 Three-stages in solving an image processing problem.

the degradation is characterized by an LSI system whose point
spread function is a circular disk followed by additive noise.
The vision model is mean squared error. The mathematical technique
is LSI least mean-square filtering. The implementation can, for
example, be done via FFT on a general purpose computer.

In the past, most research efforts have concentrated on mathematical techniques and image models. The mathematical techniques
are usually implemented on general-purpose computers. More recently as the field of image processing matures, more and more attention is turning toward special-purpose computer systems and
hardware.

In image modeling, an exciting new area is beginning to open
up. This is the modeling of image sequences (moving images).
Much work has been done in the efficient coding of image sequences.
However, the enhancement and analysis of image sequences are in
their infancy - we need good mathematical models for developing
mathematical techniques.

In mathematical techniques, the trend is to pool together ideas
and concepts from diverse fields, especially signal processing,
pattern recognition, and artificial intelligence. This is perhaps
most evident in scene analysis. Referring to Fig. 1, we see that
signal processing techniques are needed for preprocessing, pattern
recognition techniques are needed for segmentation, feature extraction, and classification, and artificial intelligence techniques are needed for structure analysis, knowledge acquisition and
representation (world model), and control structures (interaction
among the blocks).

In implementation, there are three interrelated directions.
The first is the development of general-purpose image processing
computers. Here, one needs to decide on a common set of mathematical tasks which are required for major image processing problems, and then investigate what computer architectures (MIMD, pipeline, associative processors, etc.) are suitable for each task.
It is likely that different architectures are required for different tasks. Therefore, a general-purpose image processing computer
should perhaps be reconfigurable.

The second is the development of large image data base systems.
Here, the paramount consideration is data structure and management.
There are many examples of large image data bases : LANDSAT, automated map making and updating, office automation, hospital patient
files, library archives, etc.

The third is the development of real-time hardware. Here,
the emphasis is on real-time. Applications include efficient

coding of video signals in teleconferencing, quality control in an
industrial production environment, and enhancement of military
FLIR (forward looking infrared) images. Hardware for basic operations such as linear filtering can of course be incorporated in
general-purpose image processing computers and large image data
base systems. Operations such as edge detection, median filtering,
and histograming are being implemented on chips and using CCD.

In Fig. 2, the most difficult problem is probably vision models.
Many attempts have been made, but there has been only very limited
success. It seems very difficult, if not impossible, to find
mathematical models for human vision which are both realistic and
mathematically tractable.

THEORY

A 1D APPROACH TO 2D-SIGNAL PROCESSING

H. W. Schuessler
Universitaet Erlangen-Nuernberg
Erlangen
Federal Republic of Germany

ABSTRACT: After a short review of 1D linear systems the usual method for 2D signal processing will be reviewed. Its merits and disadvantages will be treated briefly. Starting with the problem of processing a sequence of images, a new approach will be proposed, based on the concept of a multiport 1D System, to be described by state equations. In this case some specializations are appropriate, leading to certain properties of the matrices describing these systems. Some stability considerations will be presented as well. The processing of a single image will be treated as a special case.

1. ONE-DIMENSIONAL LINEAR SYSTEMS

As an introduction to the usual approach for 2D signal processing as well as to the method presented here, a short summary of the descriptions and properties of 1D sequences and systems will be given first [1].
We consider a system described by

$$y(k) = S\{u(k)\} , \tag{1.1}$$

where $\{u(k)\}$ and $\{y(k)\}$ are the input and output sequences, respectively. Here $u,y \in \mathbb{C}$ and $k \in \mathbb{Z}$. Especially, we use certain test-sequences at the input of the system.

$$\underline{\text{Step:}} \quad u(k) = \gamma_{-1}(k) = \begin{cases} 1 & k \geq 0 \\ 0 & k < 0 \end{cases} \tag{1.2}$$

Impulse: $u(k) = \gamma_0(k) = \begin{cases} 1 & k = 0 \\ 0 & k \neq 0 \end{cases}$ (1.3)

Exponential: $u(k) = U\, z_1^{\,k}$ (1.4)

with $U, z_1 \in \mathbb{C}$.

The linear system, we are interested in, can be described completely by its impulse response, the response to an impulse applied at k=κ:

$$h_0(k,\kappa) = S\{\gamma_0(k-\kappa)\} \; .$$ (1.5)

This holds for the most general case of a linear, not necessarily causal and time invariant system. As a further specialization we assume causality and an input sequence with $u(k) = 0$ $\forall\ k < 0$. Since $u(k)$ can be expressed as

$$u(k) = \sum_{\kappa=0}^{k} u(\kappa)\gamma_0(k-\kappa) \; ,$$ (1.6)

the output sequence $y(k)$ turns out to be

$$y(k) = \sum_{\kappa=0}^{k} u(\kappa)h_0(k,\kappa) \; .$$ (1.7)

With the input- and output vectors $\mathbf{u}(k)$ and $\mathbf{y}(k)$, respectively, defined as

$$\mathbf{u}(k) = [u(0), u(1), u(2), \ldots]^T$$ (1.8)

$$\mathbf{y}(k) = [y(0), y(1), y(2), \ldots]^T$$ (1.9)

this relation can be written as

$$\begin{bmatrix} y(0) \\ y(1) \\ y(2) \\ \vdots \end{bmatrix} = \begin{bmatrix} h_0(0,0) & 0 & 0 & \cdots \\ h_0(1,0) & h_0(0,1) & 0 & \cdots \\ h_0(2,0) & h_0(1,1) & h_0(0,2) & \cdots \\ \vdots & \vdots & \vdots \end{bmatrix} \cdot \begin{bmatrix} u(0) \\ u(1) \\ u(2) \\ \vdots \end{bmatrix}$$

or

$$\mathbf{y}(k) = S \cdot \mathbf{u}(k) \; .$$ (1.10)

The matrix S, consisting of values $h_0(k,\kappa)$, describes the system. Assuming further time-invariance we have

$$S\{\gamma_0(k-\kappa)\} = h_0(k-\kappa) \; \forall\ \kappa \; .$$ (1.11)

Equation (1.7) specializes to a convolution of the sequences $\{u(k)\}$ and $\{h_0(k)\}$

$$y(k) = \sum_{\kappa=0}^{k} u(\kappa)h_0(k-\kappa) = u(k) * h_0(k) \; , \qquad (1.12)$$

which in turn can be written as

$$\begin{bmatrix} y(0) \\ y(1) \\ y(2) \\ \vdots \end{bmatrix} = \begin{bmatrix} h_0(0) & 0 & 0 & \cdots \\ h_0(1) & h_0(0) & 0 & \cdots \\ h_0(2) & h_0(1) & h_0(0) & \cdots \\ \vdots & \vdots & \vdots & \vdots \end{bmatrix} \cdot \begin{bmatrix} u(0) \\ u(1) \\ u(2) \\ \vdots \end{bmatrix} \; . \qquad (1.13)$$

Thus the system matrix **S** becomes a Töplitz matrix.

In the following we assume linear, causal and time-invariant systems, which can be described by difference equations. If the system is of n-th order, its state is represented by a state vector **x**(k) with n elements. We have the state equations

$$\mathbf{x}(k+1) = \mathbf{A}\,\mathbf{x}(k) + \mathbf{B}\,u(k)$$

$$y(k) = \mathbf{C}\,\mathbf{x}(k) + d\,u(k) \quad , \qquad (1.14)$$

where the matrices have the dimensions $D[\mathbf{A}]=n{\times}n$, $D[\mathbf{B}]=n{\times}1$, $D[\mathbf{C}]=1{\times}n$. With the initial state **x**(0) = **0** the impulse response turns out to be

$$h_0(k) = d\cdot\gamma_0(k) + \mathbf{C}\,\mathbf{A}^{k-1}\cdot\mathbf{B}\gamma_{-1}(k-1) \quad . \qquad (1.15)$$

With the Z-transform we get the transfer function as an equivalent description, again for **x**(0) = **0**

$$H(z) = Z\{h_0(k)\} = \sum_{k=0}^{\infty} h_0(k)z^{-k} = d + \mathbf{C}(z\,\mathbf{I} - \mathbf{A})^{-1}\mathbf{B} \qquad (1.16)$$

$$= \frac{\displaystyle\sum_{\mu=0}^{m} b_{\mu}\,z^{\mu}}{\displaystyle\sum_{\nu=0}^{n} c_{\nu}\,z^{\nu}} = \frac{B(z)}{C(z)} = \frac{Y(z)}{U(z)} \quad . \qquad (1.17)$$

Here $Y(z) = Z\{y(k)\}$ and $U(z) = Z\{u(k)\}$ are the Z-transform of the output and input sequences of the system respectively. Finally we get a difference equation of n-th order, if we write eq. (1.17) as

$$Y(z) \cdot C(z) = B(z) \cdot U(z) \quad ,$$

or equivalently in the time domain as

$$y(k) * \{c_\nu\} = u(k) * \{b_\mu\} \quad ,$$

leading to

$$\sum_{\nu=0}^{n} c_\nu y(k+\nu) = \sum_{\mu=0}^{m} b_\mu u(k+\mu) \quad .$$

With $c_n = 1$ we have

$$y(k+n) = \sum_{\mu=0}^{m} b_\mu u(k+\mu) - \sum_{\nu=0}^{n-1} c_\nu y(k+\nu) \quad . \qquad (1.18)$$

Finally the principal types of systems and some of their structures are mentioned briefly. A recursive system is characterized by the fact, that we have $c_n=1$, but $c_\nu \neq 0$ for at least one

$\nu \in [0, n-1]$, or by stating, that

$$h_0(k) \neq 0 \qquad \text{possible} \quad \forall \, k \in \mathbb{N}_0 \quad .$$

They are called IIR (infinite-impulse-response) systems as well. Figure 1 shows the signal flow graph of the direct structure in its first form. The state variables are indicated.

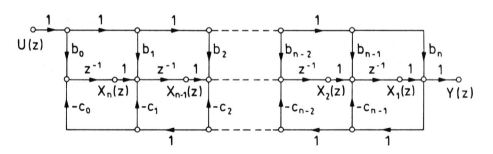

Fig. 1   The direct form of a recursive system

Another very common representation is obtained, if we write $H(z)$ as

$$H(z) = \prod_{\lambda=1}^{\ell} H_\lambda(z) \text{ with } \ell = n/2, \qquad (1.19)$$

where

$$H_\lambda(z) = \frac{b_{2\lambda} z^2 + b_{1\lambda} z + b_{0\lambda}}{z^2 + c_{1\lambda} z + c_{0\lambda}} \qquad (1.20)$$

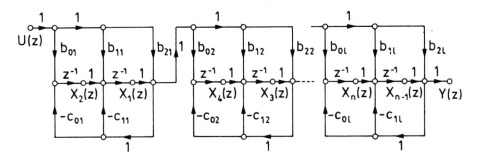

Fig. 2   The cascade form of a recursive system

is the transfer function of a subsystem of second order. Here (1.19) is the description of a cascade of blocks of second order (Fig. 2). Again the state variables are shown.

With $c_n = 1$ and $c_\nu = 0$ for $\nu \in [0, n-1]$, we get a nonrecursive system, the impulse response of which is of finite length $n+1$, i.e.,

$$h_0(k) = 0 \qquad \forall \; k \notin [0, n] \qquad ;$$

this is also called a FIR (finite-impulse-response) system. Its transfer function is

$$H(z) = \frac{1}{z^n} \sum_{\mu=0}^{n} b_\mu z^\mu = \sum_{\mu=0}^{n} h_0(\mu) z^{-\mu}$$

(1.21)

with

$$b_{n-\mu} = h_0(\mu) \qquad .$$

A possible structure is shown in Figure 3a, the so-called second form of the direct structure, specialized to the nonrecursive case. An important further specialization is obtained, if e.g., for $n = 2N$ (even)

$$h_0(\mu) = h_0(n-\mu) \qquad \mu \in [0, N-1] \; .$$

Here the system has linear phase. Figure 3b shows the structure. We note, that in all these cases the size of the data memory, i.e., the length of the state vector is equal to the degree of $C(z)$, the denominator of the transfer function.

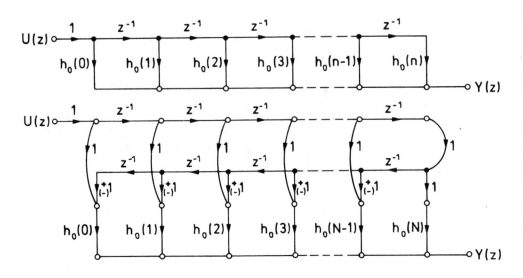

Fig. 3   The direct form of a nonrecursive system
          a) without, b) with linear phase

2.   TWO-DIMENSIONAL LINEAR SYSTEMS

The usual approach to 2D systems is obtained, if we extend the
1D concept by using two independent variables instead of one (e.g.,
[2]). Thus we are dealing with 2D sequences $\{u(\ell_1,\ell_2)\}$ and
$\{y(\ell_1,\ell_2)\}$ as input and output sequences, respectively, of a system
described by

$$y(\ell_1,\ell_2) = S\{u(\ell_1,\ell_2)\} \quad . \tag{2.1}$$

Here $u,y \in \mathbb{C}$ and $\ell_1,\ell_2 \in \mathbb{Z}$. Again test sequences are used at
the input of the system.

$$\underline{\text{Step:}} \quad u(\ell_1,\ell_2) = \gamma_{-1}(\ell_1,\ell_2) = \begin{cases} 1 & \ell_1 \text{ and } \ell_2 \geq 0 \\ 0 & \ell_1 \text{ or } \ell_2 < 0 \end{cases} \tag{2.2}$$

$$\underline{\text{Impulse:}} \quad u(\ell_1,\ell_2) = \gamma_0(\ell_1,\ell_2) = \begin{cases} 1 & \ell_1 = \ell_2 = 0 \\ 0 & \text{elsewhere} \end{cases} \tag{2.3}$$

$$\underline{\text{Exponential:}} \quad u(\ell_1,\ell_2) = z_1^{\ell_1} \cdot z_2^{\ell_2}, \; z_{1,2} \in \mathbb{C} . \tag{2.4}$$

These are examples of separable sequences defined as

$$u(\ell_1,\ell_2) = u(\ell_1) \cdot u(\ell_2) \quad . \tag{2.5}$$

While this definition of 2D sequences is a straight-forward extension of that for 1D sequences, we note, that in practical cases we have for images a finite range for the independent variables only. That means we have

$$u(\ell_1,\ell_2) \neq 0 \text{ possible only for } \quad 1 \leq \ell_1 \leq L_1$$

$$1 \leq \ell_2 \leq L_2 \ .$$

If the 2D system (2.1) is linear, it can be described in general by its impulse response

$$h_0(\ell_1,\lambda_1;\ell_2,\lambda_2) = S\{\gamma_0(\ell_1-\lambda_1;\ell_2-\lambda_2)\} \quad . \tag{2.6}$$

While in the 1D case, where usually the independent variable k is increasing with time, causality is an inherent property of a real system, this condition is not necessary in the 2D case, where we might assume, that the whole image $u(\ell_1,\ell_2)$ is known, before we start to process it. In spite of that we assume causality, i.e.

$$h_0(\ell_1,\ell_2) \equiv 0 \text{ for } \ell_1 \text{ or } \ell_2 < 0 \quad . \tag{2.7}$$

For a causal input sequence $u(\ell_1,\ell_2) \equiv 0$ for ℓ_1 or $\ell_2 < 0$ expressed as

$$u(\ell_1,\ell_2) = \sum_{\lambda_1=0}^{\ell_1} \sum_{\lambda_2=0}^{\ell_2} u(\lambda_1,\lambda_2)\gamma_0(\ell_1-\lambda_1;\ell_2-\lambda_2)$$

we get

$$y(\ell_1,\ell_2) = \sum_{\lambda_1=0}^{\infty} \sum_{\lambda_2=0}^{\infty} u(\lambda_1,\lambda_2) \cdot h_0(\ell_1,\lambda_1;\ell_2,\lambda_2) \ . \tag{2.8}$$

As a further specialization we assume shift invariance which leads to

$$y(\ell_1,\ell_2) = \sum_{\lambda_1=0}^{\ell_1} \sum_{\lambda_2=0}^{\ell_2} u(\lambda_1,\lambda_2) \cdot h_0(\ell_1-\lambda_1;\ell_2-\lambda_2) \tag{2.9}$$

$$= u(\ell_1,\ell_2) \ ** \ h_0(\ell_1,\ell_2) \quad .$$

Using the two-dimensional Z-transform we get the transfer function of the 2D linear, causal and shift invariant system

$$H(z_1,z_2) = Z_{2D}\{h_0(\ell_1,\ell_2)\} = \sum_{\ell_1=0}^{\infty} \sum_{\ell_2=0}^{\infty} h_0(\ell_1,\ell_2)z_1^{-\ell_1} z_2^{-\ell_2}. \tag{2.10}$$

Let especially $H(z_1,z_2)$ be the quotient of polynomials in z_1 and z_2

$$H(z_1,z_2) = \frac{\sum_{\mu=0}^{m_1} \sum_{\nu=0}^{m_2} b_{\mu\nu} z_1^{\mu} z_2^{\nu}}{\sum_{\mu=0}^{n_1} \sum_{\nu=0}^{n_2} c_{\mu\nu} z_1^{\mu} z_2^{\nu}} = \frac{B(z_1,z_2)}{C(z_1,z_2)} = \frac{Y(z_1,z_2)}{U(z_1,z_2)} \quad , \quad (2.11)$$

where $Y(z_1,z_2) = Z_{2D}\{y(\ell_1,\ell_2)\}$, $U(z_1,z_2) = Z_{2D}\{u(\ell_1,\ell_2)\}$.

E.g., for $m_1=m_2 = n_1=n_2 = 2$ we have

$$H(z_1,z_2) = \frac{b_{22}z_1^2 z_2^2 + b_{21}z_1^2 z_2 + b_{12}z_1 z_2^2 + \ldots + b_{10}z_1 + b_{01}z_2 + b_{00}}{c_{22}z_1^2 z_2^2 + c_{21}z_1^2 z_2 + c_{12}z_1 z_2^2 + \ldots + c_{10}z_1 + c_{01}z_2 + c_{00}}$$

$$H(z_1,z_2) = \frac{\begin{bmatrix} 1 & z_1 & z_1^2 \end{bmatrix} \begin{bmatrix} b_{00} & b_{01} & b_{02} \\ b_{10} & b_{11} & b_{12} \\ b_{20} & b_{21} & b_{22} \end{bmatrix} \begin{bmatrix} 1 \\ z_2 \\ z_2^2 \end{bmatrix}}{\begin{bmatrix} 1 & z_1 & z_1^2 \end{bmatrix} \begin{bmatrix} c_{00} & c_{01} & c_{02} \\ c_{10} & c_{11} & c_{12} \\ c_{20} & c_{21} & c_{22} \end{bmatrix} \begin{bmatrix} 1 \\ z_2 \\ z_2^2 \end{bmatrix}} = \frac{\mathbf{z}_1^T \mathbf{B} \mathbf{z}_2}{\mathbf{z}_1^T \mathbf{C} \mathbf{z}_2} \quad .$$

The actual implementation of the system can be described by a difference equation, obtained as follows. (2.11) leads to

$$Y(z_1,z_2) \cdot C(z_1,z_2) = U(z_1,z_2) \cdot B(z_1,z_2)$$

or in the space domain

$$y(\ell_1,\ell_2) ** \{c_{\mu\nu}\} = u(\ell_1,\ell_2) ** \{b_{\mu\nu}\} \quad .$$

Thus $\sum_{\mu=0}^{n_1} \sum_{\nu=0}^{n_2} c_{\mu\nu} \cdot y(\ell_1+\mu; \ell_2+\nu) = \sum_{\mu=0}^{m_1} \sum_{\nu=0}^{m_2} b_{\mu\nu} u(\ell_1+\mu; \ell_2+\nu).$

With $c_{n_1 n_2} = 1$ we have

$$y(\ell_1+n_1;\ell_2+n_2) = \sum_{\mu=0}^{m_1} \sum_{\nu=0}^{m_2} b_{\mu\nu} u(\ell_1+\mu;\ell_2+\nu) -$$

$$(2.12)$$

$$- \sum_{\substack{\mu=0 \\ (\mu,\nu)\neq(n_1,n_2)}}^{n_1} \sum_{\nu=0}^{n_2} c_{\mu\nu} y(\ell_1+\mu;\ell_2+\nu) .$$

This equation describes the recursive calculation of one output value $y(\ell_1+n_1; \ell_2+n_2)$ using the input values and known, already calculated output values. We observe, that this recursion leads to a certain direction in the processing of the image, which is not given by the nature of the problem. If, e.g. we rotate the input image by, say 90°, which is not an essential alternation, the output image will be different due to this direction of processing.

Furthermore we consider the different tpyes of systems as we did in the one-dimensional case. We have a recursive system for $c_{n_1 n_2} = 1$ and $c_{\mu\nu} \neq 0$ for at least one $\mu \in [0,n_1-1]$, $\nu \in [0,n_2-1]$.

Figure 4 illustrates, how an output value $y(\ell_1,\ell_2)$ can be calculated, if the input sequence is causal. In general initial values have to be known for the solution of the difference-equation. Usually they are assumed to be zero. We note, that the size of the data-memory is mainly determined by the size of the image.

We get the nonrecursive or FIR-case for $c_{n_1 n_2} = 1$ and $c_{\mu\nu} = 0$ for $(\mu,\nu) \neq (n_1,n_2)$. The difference equation turns out to be

$$y(\ell_1+n_1;\ell_2+n_2) = \sum_{\substack{\mu=0 \\ (\mu,\nu)\neq(n_1,n_2)}}^{n_1} \sum_{\nu=0}^{n_2} b_{\mu\nu} u(\ell_1+\mu;\ell_2+\nu) .$$

$$(2.13)$$

With

$$b_{n_1-\mu,n_2-\nu} \overset{!}{=} h_0(\mu,\nu) \overset{!}{=} h_0(\ell_1,\ell_2)$$

we have

$$h_0(\ell_1,\ell_2) \equiv 0 \quad \forall \, (\ell_1,\ell_2) \notin [0,n_1; \, 0,n_2]$$

and thus

$$y(\ell_1,\ell_2) = \sum_{\lambda_1=\ell_1}^{\ell_1-n_1} \sum_{\lambda_2=\ell_2}^{\ell_2-n_2} h_0(\ell_1-\lambda_1;\ell_2-\lambda_2)u(\lambda_1,\lambda_2) . \qquad (2.14)$$

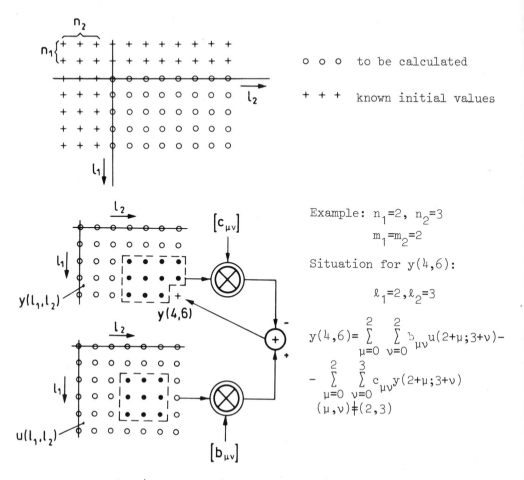

Fig. 4   Processing by a Recursive System

The transfer function becomes

$$H(z_1, z_2) = \sum_{\ell_1=0}^{n_1} \sum_{\ell_2=0}^{n_2} h_0(\ell_1, \ell_2) z_1^{-\ell_1} z_2^{-\ell_2} . \qquad (2.15)$$

Figure 5 illustrates the processing for $n_1=2$ and $n_2=3$. A system with linear phase is of special interest for image processing. We get it for example with the symmetrical impulse response

$$h_0(\ell_1, \ell_2) = h_0(n_1-\ell_1, n_2-\ell_2),$$

but there are other possibilities.

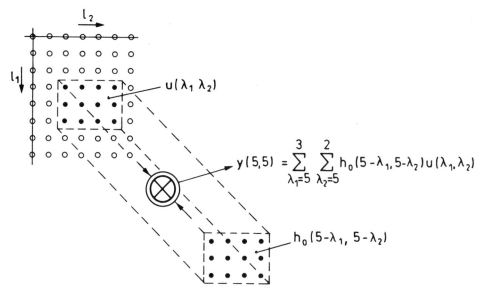

Fig. 5   Processing by a Nonrecursive System with $n_1=2$, $n_2=3$.

In summary we can compare 1D and 2D systems and their treatment as shown in Table 1. The main differences are:

1. In 1D systems we handle usually sequences of infinite length, while in 2D systems, at least if we process images, the sequences are of limited length in practical cases.

2. While in 1D systems usually (with time as independent variable) causality is a necessity, this is not the case in 2D systems, where we can assume the complete image to be given beforehand.

3. In 1D systems the direction of processing is given, this is not the case in image processing. But since the processing introduces a certain direction, a rotation of the input image, which is quite possible, will lead to another output image.

4. The size of the data memory is equal to the order of the system in the 1D case; it depends on the signal in the 2D case.

5. Due to the convolution (1.12) and (2.9) the size of the sequences at the output is always larger than that at the input, if we assume $\{u(k)\}$ and $\{u(\ell_1,\ell_2)\}$ to be of finite length. Since, in general, this assumption does not hold in the 1D case, this property effects the 2D case only.

	1D	2D
Input sequences:		
function of	a time variable k	space variables ℓ_1,ℓ_2
size	not nec. limited $k \in \mathbb{Z}$	finite $1 \leq \ell_{1,2} \leq L_{1,2}$
System: variant or invariant	in time	in space
causality	necessary	not necessary
Description by	impulse response state equations difference equation transfer function	
Types of Systems	Recursive (IIR) Nonrecursive (FIR)	
Size of the memory (Order of the System)	does not depend on the signal	depends on the signal
Size of output sequences	in general unlimited	$> L_1 \cdot L_2$

Table 1   Comparison of 1D- and 2D systems

In spite of the fact, that these main differences are usually ignored due to their possibly minor influence, another approach is of interest, which takes into account the special properties of the 2D input sequences. This approach will be introduced by considering first the more general problem of handling moving images.

3.   PROCESSING OF MOVING IMAGES

A moving image can be described by a function $u(t,\ell_1,\ell_2)$, where t is a continuous time variable $0 \leq t < \infty$ and ℓ_1,ℓ_2 are continuous space variables with $0 < \ell_{1,2} \leq L_{1,2}$. With a proper sampling in time and space we get a sequence of images $u(k,\ell_1,\ell_2)$, where now, after normalization, the variables are integers $k \in \mathbb{N}_0$, $\ell_1 = 1(1)L_1$, $\ell_2 = 1(1)L_2$. This image can be represented by a vector of length $L = L_1 \cdot L_2$

$$\mathbf{u}(k) = \begin{bmatrix} \mathbf{u}(k,\ 1) \\ \vdots \\ \mathbf{u}(k,\ \ell_1) \\ \vdots \\ \mathbf{u}(k,\ L_1) \end{bmatrix} , \qquad (3.1)$$

with $\mathbf{u}(k,\ell) = [u(k,\ell_1,1),\ u(k,\ell_1,2),\ \ldots\ u(k,\ell_1,L_2)]^T$, describing the ℓ_1th row of the image

Now we consider a system as given in Fig. 6. It handles sequences of images, taking into account 3 (in general m) input images and formerly calculated 2 (in general n) output images. We call it a system of block degree 2 (in general block degree n). It is a 1D system with an input vector $\mathbf{u}(k)$ of length L and an output $\mathbf{y}(k)$ of the same length. Thus it can be described by the state equations

$$\mathbf{x}(k+1) = \underline{\underline{A}}\ \mathbf{x}(k) + \underline{\underline{B}}\ \mathbf{u}(k)$$
$$\mathbf{y}(k) = \underline{\underline{C}}\ \mathbf{x}(k) + \underline{\underline{D}}\ \mathbf{u}(k) \qquad \mathbf{x}(k) = \begin{bmatrix} \mathbf{x}_1(k) \\ \mathbf{x}_2(k) \end{bmatrix}. \qquad (3.2)$$

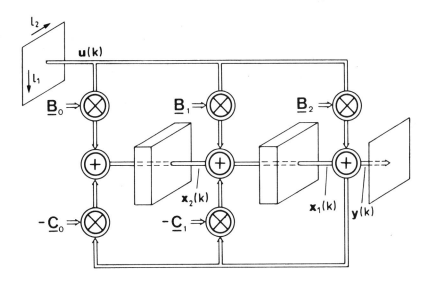

Fig. 6   Processing of Moving Images

In order to explain these expressions we consider firstly a product

$$\mathbf{v} = \underline{G}\,\mathbf{w} \quad , \tag{3.3}$$

where $\mathbf{w}$ and $\mathbf{v}$ are vectors of length $L = L_1 \cdot L_2$ representing images of size $L_1 \times L_2$ as explained in (3.1). The properties of the matrix $\underline{G}$ are choosen according to the problem at hand, which can be considered as a mapping of images. For the schematic in Fig.7 we use the following assumptions:

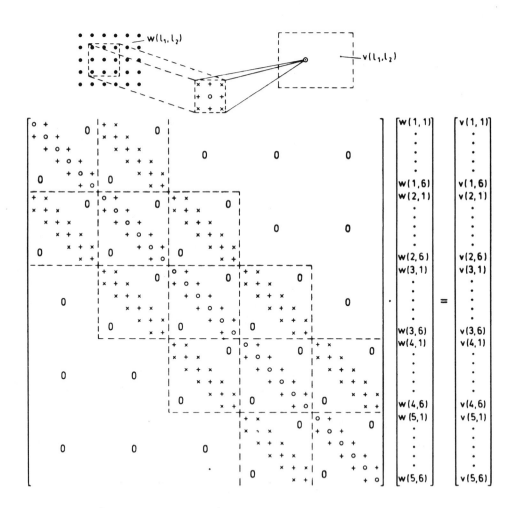

Fig. 7   Illustration of the Product $\mathbf{v} = \underline{G} \cdot \mathbf{w}$.

1. Each point of the image **v** is influenced by (at most) 9 points of the image **w**, the direct neighbors to its position.
2. The relationship is symmetrical in space.

Figure 7 illustrates the properties of $\underline{G}$ for $L_1=5$ and $L_2=6$. The signs o, +, x symbolize possibly different elements of $\underline{G}$, which are $\neq$ 0. Thus $\underline{G}$ of Fig. 7 can be written as

$$\underline{G} = \begin{bmatrix} G_0 & G_1 & 0 \text{------} 0 & 0 \\ G_1 & G_0 & G_1 & & 0 \\ 0 & G_1 & G_0 & G_1 & & 0 \\ & & G_1 & G_0 & & 0 \\ 0 & & & & & G_1 \\ 0 & 0 \text{------} 0 & G_1 & G_0 \end{bmatrix} . \qquad (3.4)$$

Here $\underline{G}$ turns out to be a $L_1 \times L_1$ tridiagonal blockmatrix, consisting of $L_1 \times L_1$ submatrices, which are tridiagonal itself, having the dimension $L_2 \times L_2$.

$$G_0 = \begin{bmatrix} g_{00} & g_{01} & 0 \text{------} 0 & 0 \\ g_{01} & g_{00} & g_{01} & & 0 \\ 0 & g_{01} & g_{00} & g_{01} & & 0 \\ & & g_{01} & g_{00} & & 0 \\ 0 & & & & & g_{01} \\ 0 & 0 \text{------} 0 & g_{01} & g_{00} \end{bmatrix} \quad ;$$

$$(3.5)$$

$$G_1 = \begin{bmatrix} g_{10} & g_{11} & 0 \text{------} 0 & 0 \\ g_{11} & g_{10} & g_{11} & & 0 \\ 0 & g_{11} & g_{10} & g_{11} & & 0 \\ & & g_{11} & g_{10} & & 0 \\ 0 & & & & & g_{11} \\ 0 & 0 \text{------} 0 & g_{11} & g_{10} \end{bmatrix} .$$

Using the notation of Fig. 7 for the elements we have $g_{00} \hat{=} o$, $g_{01} = g_{10} \hat{=} +$, $g_{11} \hat{=} x$. We note, that G_0 determines the contribution of one row in **w** to the corresponding one in **v** and G_1 the contributions of the neighbouring rows.

In the more general, but still symmetric case, where each

element of $\mathbf{v}$ is influenced by all elements of $\mathbf{w}$, $\underline{\underline{G}}$ is a $L_1 \times L_1$ symmetric block Töplitz matrix, consisting of symmetric Töplitz matrices of dimension $L_2 \times L_2$. But this general case will not be considered further.

Products of the type explained here occure at five points in Fig. 6. The matrices $\underline{\underline{A}}$, $\underline{\underline{B}}$ and $\underline{\underline{C}}$ in the state equations (3.2) consists of blockmatrices. We have

$$\underline{\underline{A}} = \begin{bmatrix} -\underline{\underline{C}}_1 & \underline{\underline{I}} \\ -\underline{\underline{C}}_0 & \underline{\underline{0}} \end{bmatrix} \quad ; \quad \underline{\underline{B}} = \begin{bmatrix} \underline{\underline{B}}_1 - \underline{\underline{C}}_1 \underline{\underline{B}}_2 \\ \underline{\underline{B}}_0 - \underline{\underline{C}}_0 \underline{\underline{B}}_2 \end{bmatrix} ; \qquad (3.6)$$

$$\underline{\underline{C}} = [\ \underline{\underline{I}} \quad \underline{\underline{0}}\ ] \quad ; \quad \underline{\underline{D}} = \underline{\underline{B}}_2 \quad .$$

An extension to a block degree n is straightforward. A generalization of the results, obtained in Section 1 for a 1D system with one input and one output yields the following describtions of the system considered here.

a) Impulse Response Matrix

$$\underline{\mathbf{h}}_0(k) = \underline{\underline{D}}\ \gamma_0(k) + \underline{\underline{C}}\ \underline{\underline{A}}^{k-1}\ \underline{\underline{B}}\gamma_{-1}(k-1) \quad . \qquad (3.7)$$

b) Matrix of Transfer Functions

$$\underline{\underline{H}}(z) = \underline{\underline{D}} + \underline{\underline{C}}\ (z\ \underline{\underline{I}} - \underline{\underline{A}})^{-1}\ \underline{\underline{B}} \quad . \qquad (3.8)$$

c) Difference Equation (Block Degree n)

$$\mathbf{y}(k+n) = \sum_{\mu=0}^{m} \underline{\underline{B}}_\mu\ \mathbf{u}(k+\mu) - \sum_{\nu=0}^{n-1} \underline{\underline{C}}_\nu\ \mathbf{y}(k+\nu) \quad . \qquad (3.9)$$

As in the Sections 2 and 3 we can distinguish between recursive and nonrecursive system, using the corresponding criterions. So we have a recursive system, if $\underline{\underline{C}}_\nu \neq 0$ for at least one ν. We get a possible structure for this case as an extension of Fig. 1, if we understand all variables as vectors, all multiplications as multiplications by matrices and all delay units as image memories. Another possible structure is obtained as a cascade of systems of block-degree 2 as shown in Fig. 6, corresponding to Fig. 2.

The nonrecursive case is characterized again by the condition $\underline{\underline{C}}_\nu = 0 \forall \nu \in [0, n-1]$. Thus the difference equation becomes

$$\mathbf{y}(k+n) = \sum_{\mu=0}^{n} \underline{\underline{B}}_\mu\ \mathbf{u}(k+\mu), \qquad (3.10)$$

where again the relation

$$\underline{\mathbf{B}}_\mu = \underline{\mathbf{h}}_0(n-\mu) \tag{3.11}$$

holds. We get a possible structure by extending the flowgraph of Fig. 3a to a system again, that handles vectors.

Since the whole approach is rather involved we consider in the following some interesting special cases only.

4. SPECIAL CASES

4.1 Nonrecursive System of Block Degree Zero

In the 1D case with a single input and output the system degenerates to a single multiplier, if the degree n is assumed to be zero. While this operation is usually not called a processing, a nonrecursive system of block degree 0 really makes sense. It is described by

$$\mathbf{y}(k) = \underline{\mathbf{B}}_0 \, \mathbf{u}(k) \stackrel{!}{=} \underline{\mathbf{B}} \, \mathbf{u}(k) \quad , \tag{4.1}$$

the simple multiplication by a matrix as explained in Section 3. We get a more general case than in (3.4), if we assume $\underline{\mathbf{B}}$ to be a general block Töplitz matrix.

$$\underline{\mathbf{B}} = \begin{bmatrix} \mathbf{B}_0 & \mathbf{B}_1 & \mathbf{B}_2 & & \cdot & \cdot & \cdot \\ \mathbf{B}_{-1} & \mathbf{B}_0 & & & & & \\ \mathbf{B}_{-2} & & & & & & \mathbf{B}_2 \\ & & & & & \mathbf{B}_0 & \mathbf{B}_1 \\ & & & \mathbf{B}_{-2} & \mathbf{B}_{-1} & \mathbf{B}_0 \end{bmatrix} \quad . \tag{4.2}$$

Here $\mathbf{B}_\lambda$ determines the contribution of row number $\ell_1+\lambda$ of the input image to the ℓ_1th row of the output image, where $-(L_1-1) \le \lambda \le (L_1-1)$. With special assumptions for the $\mathbf{B}_\lambda$ we get a more compact description. Let

$$\mathbf{B}_\lambda = b_\lambda \cdot \mathbf{B}_0 \quad , \tag{4.3}$$

then $\underline{\mathbf{B}} = \mathbf{B}' \otimes \mathbf{B}_0$ (4.4)

is the Kronecker product of $\mathbf{B}'$ and $\mathbf{B}_0$. Here $\mathbf{B}_0$ is a $L_2 \times L_2$ matrix as before, while $\mathbf{B}'$ is a $L_1 \times L_1$ Töplitz matrix given as

$$
\mathbf{B'} = \begin{bmatrix}
1 & b_1 & b_2 & & & \\
b_{-1} & 1 & & & & \\
b_{-2} & & & & & b_2 \\
& & & & 1 & b_1 \\
& & & b_{-2} & b_{-1} & 1
\end{bmatrix} . \tag{4.5}
$$

In this case the output image can be described as

$$
\mathbf{Y}(k) = \mathbf{B'}\,\mathbf{U}(k)\,\mathbf{B}_0 , \tag{4.6}
$$

where

$$
\mathbf{U}(k) = \begin{bmatrix}
\mathbf{u}(k,1)^T \\
\vdots \\
\mathbf{u}(k,\ell_1)^T \\
\vdots \\
\mathbf{u}(k,L_1)^T
\end{bmatrix} , \quad
\mathbf{Y}(k) = \begin{bmatrix}
\mathbf{y}(k,1)^T \\
\vdots \\
\mathbf{y}(k,\ell_1)^T \\
\vdots \\
\mathbf{y}(k,L_1)^T
\end{bmatrix} \tag{4.7}
$$

are matrix descriptions of the input and output image respectively.

The simple case of a general nonrecursive system of block degree 0 is in some respect related to a nonrecursive 2D system as described in Section 2. In both cases one point of the output image is a linear combination of some input points. The difference lies in the fact, that a nonrecursive 2D system yields an output image, which is larger than that at the input and shows the direction of processing, at least if we assume the system to be spatially invariant. Indeed we can get equivalence between these types of systems, if we assume the nonrecursive 2D system to be spatially variant with a properly chosen impulse response. Let us consider, e.g., the discrete approximation of the Laplacian operator, given by the impulse response

$$
h_0(\ell_1,\ell_2) = \begin{bmatrix}
0 & 1 & 0 \\
1 & -4 & 1 \\
0 & 1 & 0
\end{bmatrix} . \tag{4.8}
$$

Its application to an image of size $L_1 \times L_2$ yields certain edge-effects, since at the borders,the differentiation is done taking into account vanishing values on the frame of the image. Figure 8b shows the result for the regular pattern of Fig. 8a. In order to avoid that, we have to use other formulas for differentiation, if we approach the edges. That can be done by a multiplication of $\mathbf{u}(k)$ by an appropriately choosen matrix $\mathbf{B}$ according to (4.1) or by using a space-variant impulse response. We might use

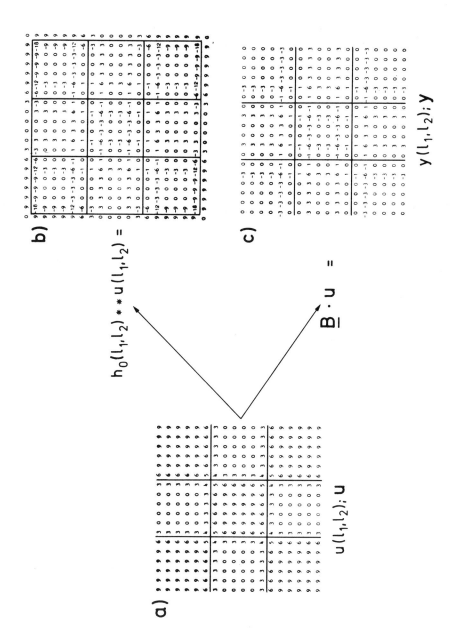

Fig. 8   Example of a Processing by the Spaceinvariant and a Spacevariant Laplacian Operator.

$$\underline{B} = \begin{bmatrix} B_0 & I & & & 0 \\ I & B_1 & & & \\ & & B_1 & & I \\ 0 & & & I & B_0 \end{bmatrix} \tag{4.9a}$$

with   (4.9b)

$$B_0 = \begin{bmatrix} -2 & 1 & & & \\ 1 & -3 & 1 & & 0 \\ & & & & \\ 0 & & -1 & -3 & 1 \\ & & & 1 & -2 \end{bmatrix} \qquad B_1 = \begin{bmatrix} -3 & 1 & & & \\ 1 & -4 & 1 & & 0 \\ & & & & \\ 0 & & -1 & -4 & 1 \\ & & & 1 & -3 \end{bmatrix} .$$

That corresponds to a processing by a 2D system with an impulse response $h_0(\ell_1,\lambda_1;\ell_2,\lambda_2)$ described as follows

$$h_0(\ell_1,\lambda_1;\ell_2,\lambda_2) = 0 \text{ for } \lambda_1,\lambda_2 \leq 0, \ \lambda_1 > L_1 \text{ or (and) } \lambda_2 > L_2$$

$$h_0(\ell_1,1;\ell_2,1) = \begin{bmatrix} -2 & 1 \\ 1 & 0 \end{bmatrix} \ ; \ h_0(\ell_1,1;\ell_2,L_2) = \begin{bmatrix} 1 & -2 \\ 0 & 1 \end{bmatrix}$$

$$h_0(\ell_1,L_1;\ell_2,1) = \begin{bmatrix} 1 & 0 \\ -2 & 1 \end{bmatrix} \ ; \ h_0(\ell_1,L_1;\ell_2,L_2) = \begin{bmatrix} 0 & 1 \\ 1 & -2 \end{bmatrix}$$

$$h_0(\ell_1,1;\ell_2,\lambda_2) = \begin{bmatrix} 1 & -3 & 1 \\ 0 & 1 & 0 \end{bmatrix} \quad 1 < \lambda_2 < L_2$$

$$h_0(\ell_1,\lambda_1;\ell_2,1) = \begin{bmatrix} 0 & 1 \\ 1 & -3 \\ 0 & 1 \end{bmatrix} \quad 1 < \lambda_1 < L_1 \tag{4.10}$$

$$h_0(\ell_1,L_1;\ell_2,\lambda_2) = \begin{bmatrix} 0 & 1 & 0 \\ 1 & -3 & 1 \end{bmatrix} \quad 1 < \lambda_2 < L_2$$

$$h_0(\ell_1,\lambda_1;\ell_2,L_2) = \begin{bmatrix} 1 & 0 \\ -3 & 1 \\ 1 & 0 \end{bmatrix} \quad 1 < \lambda_1 < L_1$$

$$h_0(\ell_1,\lambda_1;\ell_2,\lambda_2) = \begin{bmatrix} 0 & 1 & 0 \\ 1 & -4 & 1 \\ 0 & 1 & 0 \end{bmatrix} \quad \begin{array}{l} 1 < \lambda_1 < L_1 \quad \text{and} \\ 1 < \lambda_2 < L_2 \end{array} .$$

Figure 8c shows the result of a processing by a 1D system according to (4.1) with (4.9) or of using the space-variant 2D system described by (2.8) with the impulse response (4.10). It should be mentioned, that in the 2D case the description in the frequency domain by the transfer function (2.14) is not possible.

4.2   A Recursive System of Block Degree One

We investigate a system as shown in fig. 9. It is described by the difference equation

$$\mathbf{y}(k+1) = \underline{\mathbf{B}}\,\mathbf{u}(k) - \underline{\mathbf{C}}\,\mathbf{y}(k) \quad . \tag{4.11}$$

A very special form of this system with $\mathbf{B} = b\cdot\mathbf{I}$ and $\underline{\mathbf{C}} = c\cdot\mathbf{I}$ was used in [3] for the processing of TV-signals. An implementation with electro-optical means with a monitor and a TV-camera in the feedback path is described in [4], [5].

For a further investigation of (4.11) we assume especially $\underline{\mathbf{B}} = \mathbf{I}$. In order to test the stability of the system we need the $L = L_1 \cdot L_2$ eigenvalues λ_i of $\underline{\mathbf{C}}$. For stability, the condition $|\lambda_i| < 1 \forall i$ has to be satisfied. While it is rather impossible to say more for the general case, there are two simplifications, leading to a closed form solution:

If $\underline{\mathbf{C}}$ is already diagonal, the whole system is decomposed into $L = L_1 \cdot L_2$ independent systems, each with one input and one output.

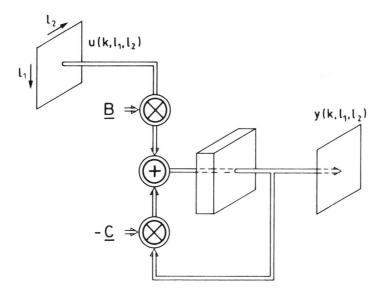

Fig. 9   A Recursive System of Block Degree One.

Of course, the elements of $\underline{C}$ are the eigenvalues, which have to satisfy the stability condition.

Secondly we assume $\underline{C}$ to be a symmetrical tridiagonal $L_1 \times L_1$ block matrix of $L_2 \times L_2$ submatrices C_0 and C_1:

$$\underline{C} = \begin{bmatrix} C_0 & C_1 & 0 & \cdots & 0 \\ C_1 & C_0 & & & 0 \\ 0 & & & & C_1 \\ \vdots & & & & \\ 0 & \cdots & 0 & C_1 & C_0 \end{bmatrix} . \tag{4.12}$$

Here the eigenvalues of $\underline{C}$ can be calculated, if C_0 and C_1 are symmetrical tridiagonal matrices themselves, i.e., if $\underline{C}$ is of the same form as $\underline{G}$ was in (3.4) or in Fig. 7. First of all, we recall, how the eigenvalues of an ordinary symmetrical tridiagonal matrix can be calculated [6]. A $N \times N$ matrix

$$C = \begin{bmatrix} c_0 & c_1 & 0 & \cdots & 0 \\ c_1 & c_0 & & & \\ 0 & & & & \\ \vdots & & & & c_0 & c_1 \\ 0 & \cdots & 0 & c_1 & c_0 \end{bmatrix} \tag{4.13}$$

can be diagonalized by a symmetrical orthogonal $N \times N$ matrix P, the elements of which are

$$P_{ik} = \sqrt{\frac{2}{N+1}} \cdot \sin \frac{ik\pi}{N+1} \qquad i,k = 1(1)N . \tag{4.14}$$

We get

$$PCP = \begin{bmatrix} \lambda_1 & 0 & \cdots & 0 \\ 0 & \lambda_2 & & \\ \vdots & & & 0 \\ 0 & \cdots & 0 & \lambda_N \end{bmatrix} , \text{ with } \lambda_\ell = c_0 + 2c_1 \cdot \cos \frac{\ell \cdot \pi}{N+1} ,$$
$$\ell = 1(1)N . \tag{4.15}$$

As an extension we use here a $L_1 \times L_1$ blockmatrix $\underline{P}_1$ for the diagonalization of $\underline{C}$. We define $\underline{P}_1$ as

$$\underline{P}_1 = P_1 \otimes I_2 \quad , \tag{4.16a}$$

where P_1 is a $L_1 \times L_1$ matrix as described by (4.14) and I_2 a $L_2 \times L_2$ identity matrix. Thus we gave for the submatrices of $\underline{P}_1$

$$P_{\ell_1 \ell_2} = \sqrt{\frac{2}{L_1+1}} \cdot \sin \frac{\ell_1 \ell_2 \pi}{L_1+1} \cdot I_2 \; ; \; \ell_1, \ell_2 = 1(1)L_1 \; . \tag{4.16b}$$

With this blockmatrix $\underline{P}_1$ we get

$$\underline{P}_1 \; \underline{C} \; \underline{P}_1 = \begin{bmatrix} \Lambda_1 & \mathbf{0} & \text{-----} & \mathbf{0} \\ \mathbf{0} & \Lambda_2 & & \\ & & \ddots & \mathbf{0} \\ \mathbf{0} & \text{----} & \mathbf{0} & \Lambda_{L_1} \end{bmatrix} \; . \tag{4.17}$$

The $\Lambda_{\ell 1}$ are the "Eigenmatrices" of dimension $L_2 \times L_2$

$$\Lambda_{\ell_1} = C_0 + 2 C_1 \cos \frac{\ell_1 \cdot \pi}{L_1+1} \qquad \ell_1 = 1(1)L_1 \; . \tag{4.18}$$

If we assume the matrices C_0 and C_1 to be tridiagonal $L_2 \times L_2$ matrices as well, defined as

$$C = \begin{bmatrix} c_{00} & c_{01} & \mathbf{0} & \text{--} & \mathbf{0} \\ c_{01} & c_{00} & & & \mathbf{0} \\ \mathbf{0} & & \ddots & & c_{01} \\ & & & & \\ \mathbf{0} & \text{---} & \mathbf{0} & c_{01} & c_{00} \end{bmatrix} , \; C_1 = \begin{bmatrix} c_{10} & c_{11} & \mathbf{0} & \text{--} & \mathbf{0} \\ c_{11} & c_{10} & & & \mathbf{0} \\ \mathbf{0} & & \ddots & & c_{11} \\ & & & & \\ \mathbf{0} & \text{--} & \mathbf{0} & c_{11} & c_{10} \end{bmatrix} . \tag{4.19}$$

These eigenmatrices become tridiagonal as well, as can be seen easily by (4.18). Thus they can be diagonalized using the same method:

$$P_2 \; \Lambda_{\ell_1} \; P_2 = D_{\ell_1} \quad , \tag{4.20a}$$

where D_{ℓ_1} is a diagonal matrix of the eigenvalues λ_{ℓ_1, ℓ_2} of Λ_{ℓ_1}

$$\lambda_{\ell_1, \ell_2} = c_{00} + 2c_{10} \cos \frac{\ell_1 \cdot \pi}{L_1+1} + 2(c_{01}+2c_{11} \cos \frac{\ell_1 \cdot \pi}{L_1+1}) \cos \frac{\ell_2 \cdot \pi}{L_2+1} \; ,$$

$$\ell_1 = 1(1)L_1 \tag{4.20b}$$
$$\ell_2 = 1(1)L_2 \quad ,$$

and $\mathbf{P}_2$ is a $L_2 \times L_2$ matrix as given by (4.14).

Finally we state, that the orthogonal matrix

$$\underline{\mathbf{M}} = \begin{bmatrix} \mathbf{P}_2 & 0 & - & - & - & 0 \\ 0 & \mathbf{P}_2 & & & \\ & & & & 0 \\ & & & & \\ 0 & - & - & - & 0 & \mathbf{P}_2 \end{bmatrix} \; \underline{\mathbf{P}}_1 = (\mathbf{I}_1 \otimes \mathbf{P}_2)(\mathbf{P}_1 \otimes \mathbf{I}_2) = \mathbf{P}_1 \otimes \mathbf{P}_2$$

$$(4.21)$$

diagonalizes $\underline{\mathbf{C}}$. The eigenvalues of $\underline{\mathbf{C}}$ are the λ_{ℓ_1, ℓ_2} as given above. Since $\ell_1 = 1(1)L_1$, $\ell_2 = 1(1)L_2$, there are $L = L_1 \cdot L_2$ eigenvalues. The system will be stable if and only if $|\lambda_{\ell_1, \ell_1}| < 1 \;\; \forall \; \ell_1, \ell_2$.

4.3 A Recursive System of Block Degree Two

We consider again the system described by fig. 6 and (3.2). According to (1.18) we write the difference equation

$$\mathbf{y}(k+2) = \underline{\mathbf{B}}_2 \, \mathbf{u}(k+2) + \underline{\mathbf{B}}_1 \, \mathbf{u}(k+1) + \mathbf{B}_0 \, \mathbf{u}(k) - \underline{\mathbf{C}}_1 \, \mathbf{y}(k+1) - \underline{\mathbf{C}}_0 \, \mathbf{y}(k).$$

$$(4.22)$$

For $\underline{\mathbf{B}}_2 = \mathbf{0}$ the matrices in the state equations (3.2) become

$$\underline{\mathbf{A}} = \begin{bmatrix} -\underline{\mathbf{C}}_1 & \mathbf{I} \\ -\underline{\mathbf{C}}_0 & \mathbf{0} \end{bmatrix} \; ; \; \underline{\mathbf{B}} = \begin{bmatrix} \mathbf{B}_1 \\ \mathbf{B}_0 \end{bmatrix} \; ; \; \underline{\mathbf{C}} = \begin{bmatrix} \mathbf{I} & \mathbf{0} \end{bmatrix} \quad \underline{\mathbf{D}} = \underline{\mathbf{0}} \; . \quad (4.23)$$

In order to check stability we need the eigenvalues of $\underline{\mathbf{A}}$. Again we assume $\underline{\mathbf{C}}_1$ and $\underline{\mathbf{C}}_0$ to be symmetric tridiagonal block matrices of the type described by (3.4). We calculate

$$\begin{bmatrix} \underline{\mathbf{P}}_1 & \mathbf{0} \\ \mathbf{0} & \underline{\mathbf{P}}_1 \end{bmatrix} \cdot \underline{\mathbf{A}} \begin{bmatrix} \underline{\mathbf{P}}_1 & \mathbf{0} \\ \mathbf{0} & \underline{\mathbf{P}}_1 \end{bmatrix} = \begin{bmatrix} -\underline{\mathbf{P}}_1 \underline{\mathbf{C}}_1 \, \underline{\mathbf{P}}_1 & \mathbf{I} \\ -\underline{\mathbf{P}}_1 \underline{\mathbf{C}}_0 \, \underline{\mathbf{P}}_1 & \mathbf{0} \end{bmatrix} \; . \quad (4.24)$$

The comparison with (4.17) shows, that

$$\underline{\mathbf{P}}_1 \underline{\mathbf{C}}_1 \, \underline{\mathbf{P}}_1 = \underline{\mathbf{\Gamma}}_1 \quad \text{and} \quad \underline{\mathbf{P}}_1 \underline{\mathbf{C}}_0 \, \underline{\mathbf{P}}_1 = \underline{\mathbf{\Gamma}}_0 \quad (4.25)$$

are diagonal matrices of eigenmatrices $\mathbf{\Lambda}_{1, \ell_1}$, $\mathbf{\Lambda}_{0, \ell_1}$, respectively.

A diagonalization of these matrices leads to a separation of the whole system into $L = L_1 \cdot L_2$ subsystems, each of second order, the stability of which can be checked easily.

4.4 Processing a Single Image

Finally we return to the problem of processing a single image. In a system for moving images it can be described as

$$\mathbf{u}(k) = \mathbf{u} \cdot \gamma_{-1}(k) \quad , \tag{4.26}$$

where the vector

$$\mathbf{u} = \begin{bmatrix} \mathbf{u}(1) \\ \vdots \\ \mathbf{u}(\ell_1) \\ \vdots \\ \mathbf{u}(L_1) \end{bmatrix}$$

describes the image. Here

$$\mathbf{u}(\ell_1) = [u(\ell_1,1),\ u(\ell_1,2)\ \ldots\ u(\ell_1,L_2)]^T \quad ,$$

is a subvector for the ℓ_1th row (see 3.1). If (4.26) is processed by a system considered here, the final value of the step response can be used as the result

$$\mathbf{y}(\infty) = \lim_{k \to \infty} S\,\{\mathbf{u} \cdot \gamma_{-1}(k)\} \ =$$
$$\mathbf{y}(\infty) = [\underline{D} + \underline{C}\,[\underline{I} - \underline{A}]^{-1}] \cdot \mathbf{u} \tag{4.27}$$
$$= S \cdot \mathbf{u} \quad .$$

Here S is a L×L matrix as $\underline{B}$ was in (4.1). A comparison of (4.27) with (4.1) shows, that a single image will be processed as in a nonrecursive system of block degree zero with an appropriately choosen matrix. As an example we calculate S for a recursive system of block degree one, as described by Fig. 9 and (4.11). Here we choose especially

$$\underline{A} = -\underline{C}\ ,\quad \underline{B} = I\ ,\quad \underline{C} = I\ ,\quad \underline{D} = \underline{0}\ , \tag{4.28}$$

where $\underline{C}$ is given by (4.12). With (4.21) we can express $\underline{C}$ as

$$\underline{C} = \underline{M}\,D_L\,\underline{M} \quad , \tag{4.29}$$

where $\mathbf{D}_L$ is the diagonal matrix of the L eigenvalues λ_{ℓ_1, ℓ_2} of $\underline{C}$, as given in (4.20). Introducing the diagonal matrix

$$\mathbf{D}_0 = [\mathbf{I} + \mathbf{D}_L]^{-1} \tag{4.30}$$

with the elements $(1 + \lambda_{\ell_1, \ell_2})^{-1}$ leads to

$$\mathbf{S} = \underline{\mathbf{M}} \cdot \mathbf{D}_0 \, \underline{\mathbf{M}} \quad . \tag{4.31}$$

Obviously the processing of the image **u** can be done in three steps:

The multiplication of **u** by $\underline{\mathbf{M}}$ can be regarded as a transformation of the image,

the multiplication by $\mathbf{D}_0$ as a processing in the transformation domain

the final multiplication by $\underline{\mathbf{M}}$ as the inverse transformation.

5. CONCLUDING REMARKS

The approach presented here provides a rather general framework, capable of describing the processing of a sequence of images or a single image having finite size.
Some systems, known from the literature turn out to be special cases of the treatment given here. Further investigations will be necessary to design systems for a certain specified processing as well as for checking the merits of this approach.

The author acknowledges the aid of Dr. Steffen, given in numerous discussions on the subject as well as the support by him and Dr. Ekstrom in proof-reading the manuscript.

REFERENCES

[1] H.W. Schüssler: Digitale Systeme zur Signalverarbeitung
     Springer-Verlag, Berlin, Heidelberg, New York 1973.

[2] R.M. Mersereau, D.E. Dudgeon: Two-Dimensional Digital Filter-
     ing. IEEE Proc. Vol. 63 (1975), pp. 610-623.

[3] R.H. Mann *et.al.*: A Digital Noise Reducer for Encoded NTSC-
     Signals. Symposium Record, Session G, International Television
     Symposium, Montreux 1977.

[4] G. Häusler, A. Lohmann: Hybrid Image Processing with Feedback.
     Optical Communications, Vol. 21 (1977), pp. 365-368.

[5] G. Häusler, M. Simon: Generation of space and time picture
     oscillations by active incoherent feedback. Optica Acta,
     Vol. 25 (1978), pp. 327-336.

[6] R. Zurmühl: Matrizen und ihre technischen Anwendungen.
     Springer-Verlag, Berlin, Göttingen, Heidelberg, 4. Auflage 1964.

DIGITAL CODING OF TELEVISION SIGNALS

H. G. Musmann
Universitaet Hannover
Hannover
Federal Republic of Germany

INTRODUCTION

Digital television coding aims at representing a television signal by as few binary symbols as possible to save channel and storage capacity in digital transmission and processing systems. In this paper, a survey of the coding concepts developed for digital television coding is presented and directions of the present research in this field are indicated.

First the concepts for intraframe coding of 5 MHz color broadcast television signals are discussed. These techniques have to meet the relatively high picture-quality requirements as recommended by the International Radio Consultative Committee. On the other hand the resulting bit rates of these coding techniques must correspond to those of the standardized PCM hierarchy to allow program exchanges between the broadcast companies.

By reducing the picture-quality requirements also the bit rate and the transmission costs can be reduced. Therefore, a special television standard with a reduced signal bandwidth of 1 MHz has been proposed for videotelephone and related applications. The coding of such low-resolution television signals is treated in the second part of this paper.

PULSE-CODE MODULATION

Digital coding of an analog television signal always
involves sampling and quantizing procedures which
theoretically produce an information loss that can but
must not lead to visible picture-quality degradations.
When a television signal is once digitized, it is
possible to apply reversible coding techniques which
exploit the signal statistics and which produce no information or quality loss.

Generally, in the first step of a coding procedure
the analog television signal is converted into a pulse-
code modulation (PCM) representation. According to the
sampling theorem a sampling frequency f_s of twice the
signal bandwidth W

$$f_s = 2W \qquad\qquad (1)$$

is required to completely determine a signal. The frequency 2W is called the Nyquist rate. For sampling
television signals sometimes a sub-Nyquist rate sampling
frequency is proposed which causes aliasing errors.
Because of the special structure of the television power

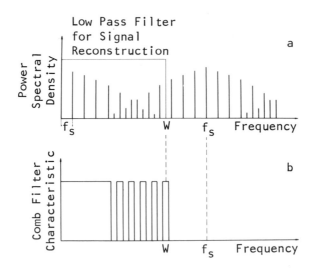

Fig.1. Power spectral density of a sampled television
signal showing interleaving spectral aliasing
components (a) and (b) a comb-filter characteristic
for reducing the aliasing errors in the reconstructed signal.

spectral density, as shown in Fig. 1, these aliasing errors can be reduced to a certain extent, if the sampling frequency is chosen to be $(2n+1)/2$ times the line frequency and a comb filter is used to eliminate the interleaving spectral aliasing components (Messerschmid, 1969). However, it should be mentioned that even with this technique, visible distortions may remain at oblique contours in a picture.

For quantizing the samples of a television signal a uniform quantizer with $K = 256$ quantizing levels is required to achieve broadcast picture quality. A reduction of the number of quantizing levels produces visible quantizing errors in form of contours, as demonstrated in Fig. 2. Thus $\log_2 K = \log_2 256 = 8$ bits are required for encoding one sample or one picture element. Depending on the signal bandwidth this so-called PCM coding yields a transmission rate of at least

$$R_{PCM} = 2W \log_2 K \qquad (2)$$

in bit per second. The PCM transmission rate is between 80 Mbit/s and 110 Mbit/s for 5-MHz color broadcast television and about 16 Mbit/s for 1-MHz videotelephone signals.

Fig.2. Television picture processed with a PCM system using 10 MHz sampling frequency and 3 bit per sample.

CODING TECHNIQUES FOR COLOR-BROADCAST TELEVISION SIGNALS

To reduce the relatively high PCM transmission rate
of color broadcast television signals, coding techniques
are investigated which can be divided into two groups,
composite and component coding methods. In composite
coding the luminance signal U_Y and the modulated subcarrier of the two chrominance signals U_{R-Y}, U_{B-Y} is
encoded as a single signal, while in component coding
U_Y, U_{R-Y}, U_{B-Y} are encoded separately. Assuming that the
component separation can be performed without introducing distortions, then component coding appears more
favorable for several reasons. Component coding allows
a uniform digital representation for color television
signals of different standards, like PAL or SECAM, and
allows a lower transmission bit rate than composite
coding.

The main coding techniques for composite and component coding are transform coding and predictive coding.
Figure 3 shows the block diagram of a transform coding
system. The matrix $\lfloor G \rfloor$ represents a block of neighboring picture elements from one line or adjacent lines.
This block is transformed with help of a one-dimensional
or two-dimensional transform $[U]$, like Hadamard or
Fourier transform. The bit-rate reduction is achieved
by quantizing the high-frequency components with fewer
bits than the low frequency components (Pratt, 1971).
Using the mean-square quantizing error as a quality

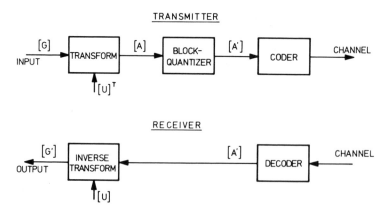

Fig.3. Block diagram of a transform coding system.

criterion, transform coding and predictive coding are
theoretically equivalent (Habibi and Hershel, 1974)
but transform coding systems are more complex. In Figure 4
the block diagram of a predictive coding system is shown.
In a predictive coding system rather than the sample
values u_N, the difference e_N between a prediction value
$\hat{u}_N$ for that sample value and the actual sample value u_N
is quantized, coded and transmitted. Different predic-
tors are required for composite and component coding.

In composite coding, the sampling frequency must be
locked to the color subcarrier frequency to get definite
phases of the subcarrier frequency at the sample points,
as demonstrated in Fig. 5, where the sampling frequency
is three times the subcarrier frequency f_{SC}. To achieve
the correct subcarrier phase for the prediction value,
three prediction algorithms must be used alternatively
(Thompson, 1975). In component coding the sampling frequencies are matched to the individual signal bandwidths
of the components. Also, three prediction algorithms are
required, one for each component, the luminance signal
U_Y and the two chrominance signals U_{R-Y}, U_{B-Y}.

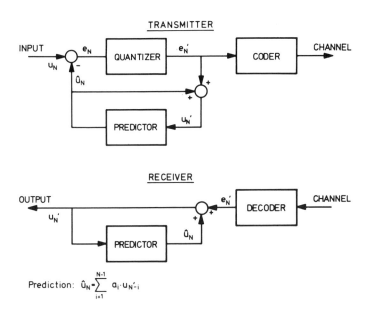

Fig.4. Block diagram of a differential pulse-code
       modulation (DPCM) system.

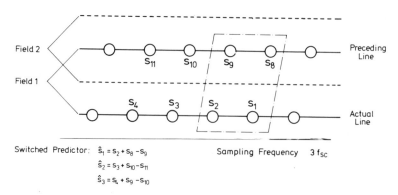

Fig.5. Position of picture elements and prediction
       algorithms for composite coding.

Figure 6 presents some prediction algorithms which have
been optimized to yield a minimum mean-square prediction
error e_N (Pirsch and Stenger, 1977). These results indicate that vertical prediction is more efficient for
the chrominance signals than for the luminance signal.

The quantizer characteristic of a DPCM-system is
matched to the masking effects of the eye. By subjective
tests it was found that with an increasing prediction

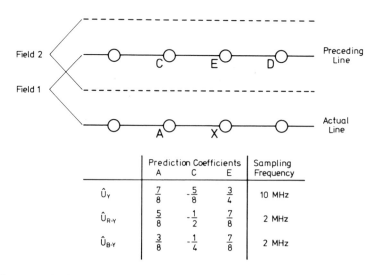

Fig.6. Position of picture elements and optimized pre-
       diction algorithms for component coding.

error a greater quantization error can be tolerated (Thoma, 1974). A more precise technique for determining the visibility threshold of quantizing errors, considering the individual influences of granular noise, edge busyness and slope overload, was developed by Pirsch (1979). From these results an optimized quantizing characteristic with a minimum number of quantizing levels can be obtained using the design procedure of Sharma and Netravali (1977). Figure 7 demonstrates, that 20 representative quantizing levels are required for quantizing the luminance signal U_Y.

A further improvement of the quantizer can be achieved by using an adaptive quantizer which exploits effects of spatial masking. Based on the masking model of Netravali and Prasada (1977) also Pirsch (1979) measured the visibility threshold of quantizing errors as a function of the luminance activity A surrounding an actual picture element to be quantized, see Fig. 8.

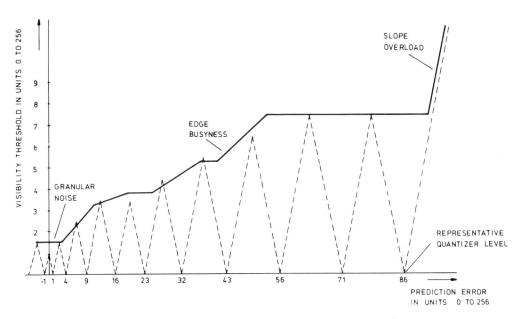

Fig.7. Visibility threshold for granular noise, edge busyness and slope overload produced by a DPCM quantizer versus magnitude of the prediction error. The dashed zig-zag line represents the quantization error of an optimized quantizer derived from the threshold function.

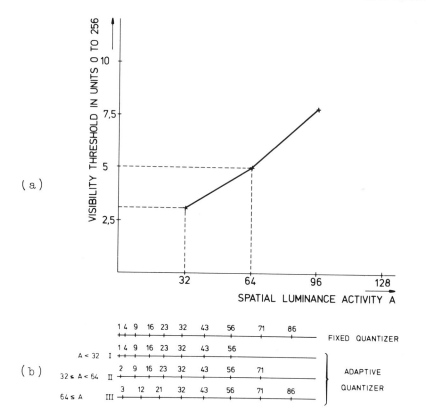

Fig.8. (a) Visibility threshold for the quantizing error
       versus luminance activity A. (b) Three quantizing
       characteristics derived from the threshold func-
       tion (a) for an adaptive quantizer.

Depending on the luminance activity, one out of three
quantizing characteristics is switched on. Each quantizing characteristic has 16 quantizing levels and is derived from the visibility threshold function assuming
a certain minimum luminance activity A. In this way,
the adaptive quantizer allows to encode more than 16
quantizing levels using 4 bits per sample without ambigu-
ity. Figure 9 demonstrates the picture quality improvement achieved with this technique for a critical test
picture. In the case of natural pictures, no picture-
quality degradations could be observed when the 4-bit
DPCM picture was compared to the original 8-bit PCM
picture. For these subjective tests, a sampling frequency
of 8.8 MHz was applied.

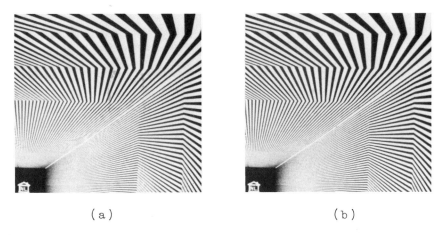

(a) (b)

Fig.9. Test picture processed with a 4-bit DPCM system
using a fixed quantizer (a) and an adaptive
quantizer (b).

The discussed coding techniques for high quality broadcast television signals lead to the sampling and quantizing configurations shown in Figures 10 and 11.
For composite coding a transmission bit rate between
53 Mbit/s and 67 Mbit/s is required. In the case of
component coding, only the second solution, shown in
Fig. 11, provides broadcast picture quality. Using this
technique a transmission rate of 34 Mbit/s corresponding
to that of the third stage of the PCM hierarchy is required. The lower bit rate of component coding mainly
results from a better, individual matching of the

CODING TECHN.	SAMPL. FREQUENCY	QUANTIZATION	BITRATE
2D DPCM	$3\,f_{SC}$	5 bit/pel	67Mbit/s
1D DPCM	$2\,f_{SC}$	6 bit/pel	53Mbit/s

2D – Two-dimensional predictor.
1D – One-dimensional predictor.

Fig.10. Sampling and quantizing configurations
for composite coding.

```
BIT RATE 34,368 Mbit/s
NUMBER OF BITS PER FRAME    : 1374720
BITS FOR SYNCHRONISATION    :   -256
BITS FOR SOUND SIGNAL       :  25000
─────────────────────────────────────
CODING BITS PER FRAME       : 1349464
CODING BITS PER LINE        :    2346
```

	U_Y	U_{R-Y}, U_{B-Y} *	Redundancy for Error Control
1. SOLUTION	$f_S = 10\,MHz$ 4 bit per pel	$f_S = 2{,}0\,MHz$ 2 bit per pel	58 bit
2. SOLUTION	$f_S = 8{,}8\,MHz$ 4 bit per pel	$f_S = 2{,}2\,MHz$ 4 bit per pel	46 bit

* Line alternating transmission of U_{R-Y} and U_{B-Y}

Fig.11. Sampling and quantizing configurations for component coding.

sampling and quantizing configurations of each component to the visual perception of the human observer.

CODING TECHNIQUES FOR LOW-RESOLUTION MONOCHROME TELEVISION SIGNALS

For special applications like videotelephone and video-conferencing television systems with a reduced picture resolution corresponding to 1 MHz bandwidth have been proposed. The PCM representation of this videotelephone signal requires a 16 Mbit/s transmission rate. Several so-called frame replenishment coding techniques (Limb, Pease and Walsh, 1974), (Haskell and Schmidt, 1975) have been published which reduce this bit rate down to about 2.0 Mbit/s providing excellent picture quality. Nevertheless, transmitting and exchanging 2 Mbit/s video-telephone signals is still a financial problem and would require a special network and special exchange systems. With the expansion of the PCM technique in the telephone network, however, transmission channels and exchange systems for digital speech signals of 64 kbit/s will be available in the near future. The question arises, whether such a digital channel can be used for transmitting a videotelephone signal and what picture quality can be obtained. Here first results are presented which have been obtained with a frame replenishment coder and

additional temporal filtering that exploits the temporal
low-pass characteristic of the eye for a further reduction of the bit rate (Musmann and Klie, 1979).

Figure 12 shows a block diagram of the complete transmission system. Broadcast television equipment is used
for the camera and for the monitor. Standard converters
convert the 625-line television signal into a 313-line
videotelephone format and vice versa.

The 64-kbit/s codec was simulated on a computer
and consists of two stages. The first stage includes
the temporal filter combined with adaptive sampling
techniques. The second stage consists of a frame replenishment coder. The block diagram in Fig. 13 indicates the two stages of the 64-kbit/s codec. The first
stage includes the temporal filter and temporal sub-
sampler, an adaptive spatial sampler and a movement
detector. In this stage, the integrating effect of the
eye is exploited which causes the known unsharp perception of moving objects. The temporal filter as shown in
Figure 14 is a low-pass filter that simulates the integrating effect of the nerve cells on the retina of the
eye. This filter also reduces the temporal bandwidth
of the television signal and allows to reduce the frame
frequency by a factor of two. Investigations have shown
that in addition to the temporal filtering this filter

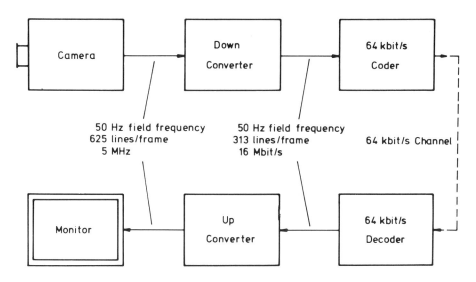

Fig.12. Block diagram of the television transmission
        system.

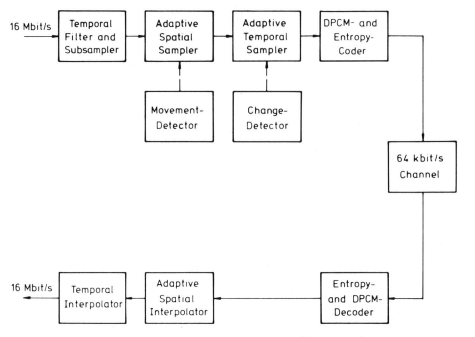

Fig.13. Block diagram of the 64 kbit/s-codec

also reduces the spatial bandwidth of moving objects.
The bandwidth decreases with increasing velocity of the
objects, see Fig. 15. Therefore, by adapting the spatial
sampling rate to the actual bandwidth of the moving objects the number of picture elements to be transmitted
can be reduced. For exploiting this effect a frame must
be clustered into moving and non moving objects with
help of a movement detector. Three blocks in Fig. 13
represent the conditional frame replenishment coder of
the second stage. A change detector detects those picture
elements which have changed from frame to frame. Only
the changed picture elements are selected by the adaptive temporal sampler and are transmitted to the receiver. Figure 16 demonstrates the segmentation of a frame
into moving objects and changed picture elements. For
encoding these picture elements a differential pulse-
code modulation (DPCM) and entropy coding is applied.

     In the decoder, the transmitted picture elements
are reconstructed. Unchanged picture elements are taken
out of a frame memory which stores the previous frame.
Picture elements eliminated by adaptive spatial sampling
or by temporal subsampling are interpolated.

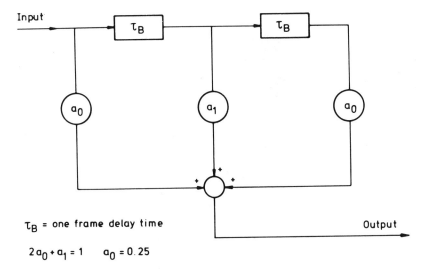

τ_B = one frame delay time

$2a_0 + a_1 = 1$ $a_0 = 0.25$

Fig.14. Block diagram of the temporal filter.

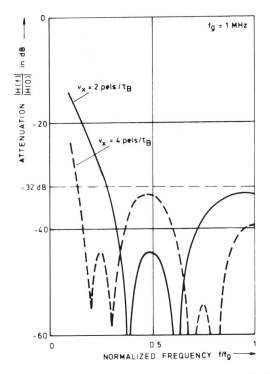

Fig.15. Spatial frequency response of the temporal
        filter for objects moving with a speed of
        2 and 4 pels per frame.

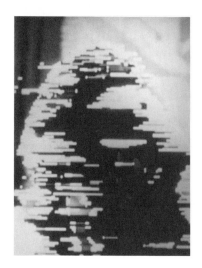

Fig.16. Moving objects and changed parts of a frame.
         Moving objects are indicated by black and
         changed parts by black and white picture
         elements.

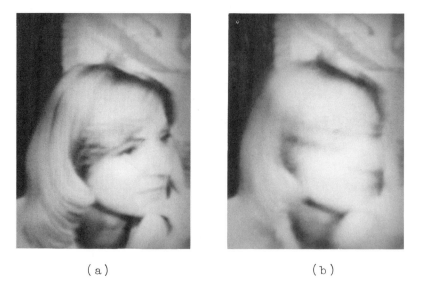

              (a)   (b)

Fig.17. One processed frame out of a sequence showing
         a person with moderate movement (a) and rapid
         movement (b).

A temporal filter combined with a reduced frame frequency of 12.5 Hz and adaptive spatial sampling of moving objects reduces the bit rate by a factor of about 8 in addition to frame-replenishment coding. The resulting bit rate varies between 40 kbit/s and 250 kbit/s depending on the amount of movement in the scene if no buffer control is applied. With help of the buffer control, the bit rate can be fixed to 64 kbit/s.

The application of temporal filtering combined with reduced frame frequency and adaptive spatial sampling in addition to frame replenishment coding improves the picture quality of scenes with slowly and moderately moving objects. For such scenes an acceptable picture quality has been obtained. However, rapidly moving objects are blurred, as shown in Fig. 17.

REFERENCES

Habibi, A., and Hershel, R.S. (1974). IEEE Trans. Commun. com-22, 692-696.

Haskell, B.G., and Schmidt, R.L. (1975). BSTJ, Vol.54, 1475-1495

Limb, J.O., Pease, R.F.W., and Walsh, K.A. (1974) BSTJ, Vol.53, 1137-1173.

Messerschmid, U. (1969). Nachrichtentech. Z., Vol.19, 515-521.

Musmann, H.G., and Klie, J. (1979). Int. Conf. Commun. 1979.

Netravali, A.N., and Prasada, B. (1977). Proc. IEEE 65, 536-548.

Pirsch, P., and Stenger, L. (1977). Acta Electronica, 19, 277-287.

Pirsch, P. (1979). Dissertation at the University of Hannover.

Pratt, W.K. (1971). IEEE Trans. Commun. com-19, 980-992.

Sharma, D.K., and Netravali, A.N. (1977). IEEE Trans. Commun. com-25, 1267-1274.

Thoma, W. (1974). Proc. Int. Zürich Sem. Digital Commun.
1974, pp. C3(1)-C3(7).

Thompson, J.E. (1975). Proc. Int. Conf. Digital Satellite
Commun., 3rd, 315-321.

DIGITAL IMAGE ANALYSIS

H. Niemann
Universitaet Erlangen-Nuernberg
Erlangen
Federal Republic of Germany

1. INTRODUCTION

Pattern classification and analysis has been a field of research for about 25 years. Some recent books in this field are [1.1-1.12].

By a "pattern" we mean some function $\underline{f}(\underline{x})$, where $\underline{f}$ and $\underline{x}$ are vectors of appropriate dimension. In the special case, where images are considered, $\underline{x}$ consists of the two coordinates of some reference system, i.e. $\underline{x} = (x_1, x_2)_t$ or $\underline{x} = (x,y)_t$. The subscript t denotes the transpose of a vector. According to the type of images, $\underline{f}$ will have a different number of components. A grey level image is represented by a scalar function $f(x,y)$, where f is a measure of the grey level at point (x,y). A color image is represented by three functions $f_r(x,y)$, $f_g(x,y)$, $f_b(x,y)$, where f_r, f_g, f_b are measures of image intensity in the channels red, green, blue. A multispectral image is represented by several functions $f_\nu(x,y)$, $\nu = 1,\ldots,M$, where f_ν is a measure of image intensity in the ν-th spectral channel and M may typically have values between four and eleven. Emphasis will be on digital methods for image analysis. In this case an image is sampled at discrete points and only these points are considered further. An analog picture is thus transformed to a discrete image array or picture matrix $\underline{f}$ yielding

$$f(x,y) \longrightarrow f(x_o + i\Delta x, y_o + j\Delta y) = f_{ij} \quad ; \quad i,j = o,1,\ldots,m-1.$$

$$[\![f_{ij}]\!] = \underline{f}$$

In order to process an image the m^2 elements of the picture matrix have to be processed. The number m depends on the details to be resolved and may be obtained from the well-known sampling theorem.

Common values are m between 256 and 2048. We shall use image and
image array or picture matrix interchangeably.

By "image analysis" we mean that an image is described by simpler
constituents and their relations to each other. Simple constituents
(or picture primitives) will depend on the particular problem to be
solved. For instance, if electrical circuit diagrams are to be analyzed, simple constituents may be resistors and transistors; if medical radiographic images are to be analyzed, simple constituents
may be contours of the ribs, an outline of the heart or an area of a
tumor. Relations between constituents may, for instance, be electrical
connections in the case of circuit diagrams or relative and absolute
location in the case of radiographs. The result of analysis is, in
general, a description $^{\rho}B$ of an image $^{\rho}\underline{f}(\underline{x})$. The index ρ denotes a
particular image out of a series of images. The description should
contain the relevant information and this may vary considerably according to the problem to be solved. Thus, $^{\rho}B$ may contain the following information:
1. A complete symbolic description of an image based on simpler constituents and their relations.
2. A list of some interesting objects within the image.
3. An indication about changes which occurred between successive re-
   cordings of images.
4. A classification of an image as belonging to some specified class.
It seems appropriate to point out that classification of a fairly
simple pattern, like a standardized character or a seperately spoken
word, is different from classification of a fairly complex image,
like a chest radiograph. Generally, by "classification" we mean that
a pattern as a whole is independently classified as belonging to one
class Ω_{λ} out of k possible classes Ω_{κ}, $\kappa = 1,...,k$. A standardized
character may be treated as one entity, out of which a feature vector
is extracted and classified. This approach would not be feasible for
complex images, for instance for deciding whether a chest radiograph
is normal or abnormal. In the latter case the radiograph will be analyzed in the sense of extracting simpler constituents; these are used
to classify the whole image. It is noted that a symbolic description
of an image contains the most extensive information gathered during
analysis whereas information is compressed to its highest degree by
classifying an image.

The interest in and the need for automatic methods of image analysis have two main reasons. First, the amount of images which should
be analyzed is enormous, and second, it is in many cases difficult to
get reproducible results. Two examples may help to appreciate the
amount of information which is gathered in pictorial form. It is mentioned in [1.10] that in the USA about 650 million medical radiographs
are taken per year. If every picture is scanned with the fairly coarse
resolution of $256 \cdot 256$ rasterpoints with 8 bits of grey level this
gives about $3,4 \cdot 10^{14}$ bits per year or about 10^{7} bits per second.
In [1.11] it is pointed out that a tactical system should process

about 10^6 resolution elements per second if a 100 x 100 km^2 area is
to be monitored. The need for reproducible results is particularly
obvious for radiographs; according to [1.10] about 30% of anomalies
remain undetected by routine diagnosis.

The main applications of image analysis are:
1. medical diagnosis,
2. remotely sensed images,
3. industrial applications, and
4. various others (like bubble chamber photographs, military images,
   and fingerprints)

Images in medical diagnosis are usually grey level pictures. Radiographic images have already been mentioned; another broad class are
cytologic images, i.e., microscopic images of chromosomes, blood cells
or cervical smears. Remotely sensed images usually are multispectral
scanner images with four to eleven spectral channels. They are recorded by satellite or airplane and yield information concerning agri-
culture and forestry (such as crop type or insect damage), geology
(such as minerals detection), hydrology (such as water resources),
geography (such as urban planning) and environment (such as air pollution). Industrial applications are usually in the area of automating production processes or performing quality control. Large efforts have been made in hand-eye devices, i.e., a hydraulically operated
arm which is controlled by a TV camera. These main applications are
discussed in great detail in [1.5,1.7]. Methods and systems for image
analysis are always developed for a certain application, as mentioned
above. Presently it is not possible and certainly uneconomical to
obtain a universal system which would be suitable for any application.
We shall denote a particular well-defined application a "field of
problems" Ω. A field of problems is, for instance, analysis of multispectral scanner data, obtained in 11 channels at an altitude of
2000 m and a resolution of 2048 points per line, in order to obtain
tree types, their location and number. A field of problems Ω is defined by the set of images belonging to the particular application

$$\Omega = \{{}^{\rho}\underline{f}(\underline{x})\,|\,\rho = 1,2,\dots\}$$

Image analysis requires three basic assumptions:
1. For a particular field of problems Ω a representative sample of
   images $\omega c \Omega$ is available in order to gather information about the
   task.
2. An image consists of simpler constituents which have certain relations among each other. It may be decomposed into these constituents.
3. An image belonging to a certain field of problems Ω has a certain
   structure. This implies that not any arrangement of simpler constituents yields a well-formed image; it also implies that many
   images ${}^{\rho}\underline{f} \in \Omega$ may be described with relatively few constituents.

The flow of pictorial data in a system for image analysis is shown
in Fig. 1, according to [1.12]. The numbers in the blocks of Fig. 1

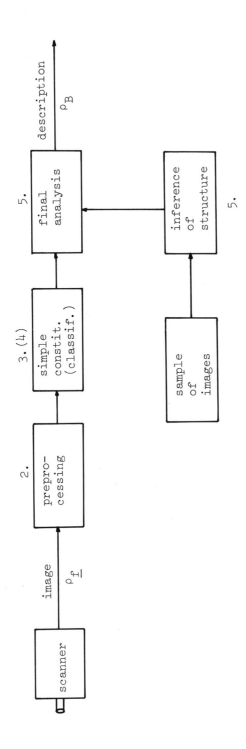

Fig. 1. Flow of pictorial data in an image-analysis system [1.12].

refer to the sections of this paper where the topics are discussed.
Flow of data should be distinguished from flow of control. In a
strictly bottom-up approach to image analysis flow of control is from
the image $^\rho f(\underline{x})$ to the description $^\rho B$. In a strictly top-down approach
flow of control would be from hypotheses about the structure to tests
of these hypotheses in the image. However, independent of the organization of the analysis process flow of pictorial data will usually
be in the sense of Fig.1. As indicated in Fig.1, devices for scanning
an image will not be discussed in this paper. Preprocessing and inference of structure will be discussed shortly, posing the main emphasis on extraction of simpler constituents and the process of analysis.

2. PREPROCESSING

The first step in image analysis is to scan the image $\underline{f}(\underline{x})$ in
order to get a picture matrix $\underline{f}$. It is assumed that no relevant information is lost during scanning. However, it may be advantageous
to preprocess $\underline{f}$ in order to reduce distortions and noise. Informally,
we mean by "preprocessing" some transformations of the scanned image
in order to get another image which is expected to facilitate further
processing and to yield improved results of analysis. Some common
techniques will be discussed in the following. The order of discussion should not suggest that the various techniques should be applied
in this order.

Coding
Image coding is used in order to remove redundancy and to efficiently store pictorial data. These problems will not be discussed
here; instead, the reader is referred to [2.1,2.2].

Filtering
Linear and homomorphic filtering is a common and powerful method
to improve image quality; unfortunately, the latter usually has to
be judged subjectively. An image $\underline{f}$ is subjected to a transformation T
to yield another image

$$\underline{h} = T\{\underline{f}\} \ . \tag{1}$$

If T is a linear shift-invariant transformation with impulse response
$\underline{g}$ and F, G, H are Fourier transforms of $\underline{f}$, $\underline{g}$, $\underline{h}$, the effect of the
transformation is described by

$$H = F \cdot G \tag{2}$$

which may conveniently be realized by digital techniques. For details
the reader is referred to [2.3,2.4]. Realization in analog form by
optical methods is also possible as described in [2.5,2.6]. Use of
polynomial transforms is developed in [2.15].

Filtering has been applied to remove high-frequency noise in images which, for instance, may result from shot noise of image dissector scanners. It is also possible to enhance certain properties of an image; for instance, high pass filtering will enhance curves and edges in the image. Inverse filtering is used in image restoration if image distortion results from some linear operation. If the distorted image H and the distortion operation G are known the original image F may be obtained from equ. (2) by solving for F. Examples are distortions by relative motion of object and film or by atmospheric turbulences. If in addition to the distortion the distorted image is influenced by additive noise, restoration may be accomplished by Wiener filtering. This technique yields the best estimate $\hat{f}$, in the sense of least mean square error, of the original f. Further details may be obtained from reference [1.6].

Grey level scaling

The grey tones of a recorded image f may deviate from the original object due to scanner characteristics, film exposure, or other reasons. Also, the subjective impression of an image may be better if the grey tones are suitably altered. In digital processing the grey tones of an image are quantized to L discrete grey levels or intervals. A common value is L = 256.

The simplest method is to map the grey tones linearly to the L discrete grey levels. Invariance to linear scaling and translation of grey tones is obtained, if the interval (l_{min}, l_{max}) of lightest to darkest grey is linearly mapped to the L discrete levels. For radiographic images a logarithmic conversion was found to be appropriate [2,7]. Film transparency $\tau(x,y)$ at point (x,y) is given by

$$\tau(x,y) = \text{local average } \{I_t(x,y) / I_i(x,y)\} \quad , \tag{3}$$

where I_t is transmitted intensity and I_i is incident intensity. Photographic density is

$$D(x,y) = \log (I_i / I_t) \quad .$$

This density is mapped to L levels. In any case, the various discrete grey levels will occur with a certain frequency which may be obtained from the grey level histogram. Modification of this histogram is a further possibility to influence grey levels. A common technique is to apply a transformation which yields an approximately constant grey level histogram. From subjective judgement, a hyperbolic histogram may be of advantage [2.8].

If f_r, f_g, f_b are the "grey" values of the red, green, blue channel of a color image, a widely used transformation is [2.9]

$$\left. \begin{array}{l} r = \dfrac{f_r}{f_r + f_g + f_b} \quad , \qquad g = \dfrac{f_g}{f_r + f_g + f_b} \, , \\[3mm] y = \alpha_1 f_r + \alpha_2 f_g + \alpha_3 f_b \end{array} \right\} \tag{4}$$

This means that the original color channels are transformed to normalized red r and green g and to total intensity y. A processing
technique based on r and g will be insensitive to intensity changes
and depend only on the "true" color.

Remote sensing of images in several spectral channels is treated
as follows. The intensity f received at the sensor depends on the intensity I_s of the source, its reflectivity ρ, the atmospheric trans-
missivity α and the background intensity β according to

$$f = \alpha \rho I_s + \beta \quad .$$

The interesting value is the reflectivity ρ. In general, all values
will depend on the spectral channel. If β is negligible and two neighbouring spectral channels, e.g. 1 and 2, are considered it is possible
to obtain the ratio

$$\frac{\rho_1}{\rho_2} = \frac{f_1}{f_2} \cdot \frac{\alpha_2 I_{s2}}{\alpha_1 I_{s1}} \quad . \tag{5}$$

If $\alpha_1 I_{s1} \approx \alpha_2 I_{s2}$, this reduces to

$$\rho_1/\rho_2 \approx f_1/f_2 \quad .$$

If an M-channel multispectral scanner yields the image

$$\underline{f}(x,y) = (f_1(x,y), f_2(x,y), \ldots, f_M(x,y))_t \quad ,$$

a normalized (M-1) component image would be

$$\underline{f}_n(x,y) = (f_1(x,y)/f_2(x,y), f_2(x,y)/f_3(x,y), \ldots, f_{M-1}(x,y)/f_M(x,y))_t$$

Other normalizing procedures may be found in [2.10].

Image normalization

Recording of an image may introduce various geometric distortions.
Also, if a larger area is subdivided and recorded in different images,
the images have to be aligned in order so that corresponding points will
match. The process of reducing geometric distortions and/or aligning
images is called image normalization. Geometric distortions may result,
for instance, from earth curvature in remotely sensed images, from
different viewing angles and altitudes, or from pincushion distortion
of CRTs.

A common normalization procedure is to obtain the grey level
$f'(x',y')$ of the transformed image as the grey level $f(x,y)$ of the
original image, where the coordinates are related by

$$x = \phi_1(x',y') \quad ,$$
$$y = \phi_2(x',y') \quad . \tag{6}$$

The functions ϕ_1, ϕ_2 may be chosen to yield projective transformations
[1.4]

$$x = \frac{a_1 x' + b_1 y' + c_1}{a_o x' + b_o y' + c_o} \qquad\qquad y = \frac{a_2 x' + b_2 y' + c_2}{a_o x' + b_o y' + c_o} \qquad (7)$$

or polynomial transformations

$$x = \sum_{\mu=o}^{p} \sum_{\nu=o}^{p-\mu} a_{\mu\nu}\, x^{\mu}\, y^{\nu}$$

$$\qquad\qquad (8)$$

$$y = \sum_{\mu=o}^{p} \sum_{\nu=o}^{p-\mu} b_{\mu\nu}\, x^{\mu}\, y^{\nu} \quad .$$

The affine transformation which takes parallel lines to parallel
lines is a special case of equ. (7) with $a_o = b_o = o$. The parameters
of the above transformations are obtained from test images or reference points. Various methods are discussed in [2.12, 2.13].

Pseudo colors

It is often desirable to have a color image of an $M > 3$ multispectral image. If the spectral channels are adjacent, information
in channel ν and $\nu + 1$ will be highly redundant. Therefore, most
of the information contained in the M channels should be representable
by three transformed channels. These three channels are assigned to
red, green, blue of a color display to yield an image with pseudo
colors. The three transformed channels are obtained by Karhunen-Loeve
or principal axis transform of the $M > 3$ channels. For details of
this transform see, e.g. [1.3, 1.4]. Another example of the use of
pseudo colors are functional images in diagnostic of series of radiographs [2.11].

Reduction of resolution

An image sampled with $m \cdot m$ points is sometimes reduced to a
lower resolution of $\frac{m}{2} \cdot \frac{m}{2}$, or $\frac{m}{4} \cdot \frac{m}{4}$, ... points. This is done
because the lower resolution may be sufficient [1.10] or because a
rough analysis, outlining only the interesting parts of the image,
is performed first and a detailed analysis of these interesting parts
is done later [2.14]. Usually the reduction is done by averaging over
a $2 \cdot 2$ or $4 \cdot 4$, and so on, neighbourhood. The neighbourhood is replaced by one point whose grey level is the average of the points in
the neighbourhood.

3. SIMPLE CONSTITUENTS

As stated in the introduction it is an important step in image
analysis to extract simpler constituents (or picture primitives) from
an image in order to describe its interesting aspects. This means
that an image is segmented or broken into parts. There are two basic
approaches to this task, which support each other.

The first approach depends on the observation that in many cases contour lines provide important information about an object. For instance, to classify an object as an airplane or a car it is sufficient to know the contours. Therefore, much effort has been taken to extract contours. Furthermore, reduction of grey level images to contour lines results in a significant information reduction and should facilitate further processing.

The second approach depends on the fact that contours alone are not sufficient in several cases. For instance, in an aerial photo the contours (or border line) of a lake or a forest provide little information about the object. If a car is reduced to its contours the color of the car, which may be important in certain applications, is lost. In these cases it is necessary to extract regions from a picture which are homogeneous in some sense; for instance, in the sense that the whole region consists of trees or is yellow.

Finally, regions may have a fairly complicated structure to which we shall refer to as its texture. For instance, pebbles at a beach will give this region a quite different appearence than fine sand. It would be difficult, if not impossible, to characterize this texture by contour lines; it would also be difficult to use some simple criterion of homogeneity like grey level or color. Because of the inherent difficulties a seperate subsection is devoted to texture. However, it should be remembered that extracting texture amounts to finding a region which is homogeneous in the sense of its textural properties; clearly, this region may be completely inhomogeneous in the sense of constant grey level.

In this section we shall be concerned only with the isolation or extraction of simple constituents. Methods to attach meaning to these constituents, i.e., to name or to classify them, will be discussed in the next sections.

3.1 Contours

The fundamental aspect of a contour is a change of grey level. Contour extraction, therefore, always requires a method to detect such changes. The fundamental difficulty with contour extraction results from the fact that noise is superimposed on any image. There is no perfect or ideal contour in a real world image. Rather than this, contours are obscured by noise, and noise may provide grey level changes in fairly homogeneous regions. To overcome these difficulties, contour extraction is done in at least two steps (we assume that any preprocessing has been performed). In the first step, points are located where grey level changes occur. In the second step, one tries to link a subset of these points by a straight or curved line. The second step usually is further subdivided.

Grey level changes

First, some methods to detect grey level changes are discussed. Surveys are given in [3.1, 3.26].

The straightforward method is to differentiate an image and to look for contours where the derivative takes high values. In digital processing this is done by computing appropriate differences. Let the grey level image be $f(x,y)$ with picture matrix $\underline{f} = [f_{ij}]$ and the differenced image be $\underline{h} = [h_{ij}]$. One of the first operators to obtain $\underline{h}$ was [3.2]

$$f_x = f_{ij} - f_{i+1,j+1} \quad ; \quad f_y = f_{i,j+1} - f_{i+1,j}$$

$$h_{ij} = \sqrt{f_x^2 + f_y^2} \quad \text{or also} \quad h_{ij} = |f_x| + |f_y| \quad . \tag{9}$$

In the above equation indices run as shown in Fig.2. If necessary, directional information is available from f_x and f_y. Better insensitivity to noise may be obtained by taking more points to compute h_{ij}. An example is [3.3]

$$f_x = (f_{i+1,j+1} + 2f_{i+1,j} + f_{i+1,j-1}) - (f_{i-1,j-1} + 2f_{i-1,j} + f_{i-1,j+1})$$

$$f_y = (f_{i-1,j+1} + 2f_{i,j+1} + f_{i+1,j+1}) - (f_{i+1,j-1} + 2f_{i,j-1} + f_{i-1,j-1}) \tag{10}$$

The meaning becomes clear if equ.(10) is examined with Fig.2. The value h_{ij} is obtained as in equ.(9). Immunity to noise may be obtained by averaging over a sufficiently large number of points. This process may be considered as a filtering operation, particularly as low pass filtering. Combining low pass filters and differencing methods to one linear operator obviously allows to generate many other operators. In a particular problem it will be necessary to experimentally select the operator which is suited best.

Reducing noise by low pass filtering also reduces the sharpness of contours. A sharp detection of grey level changes in the presence of noise is accomplished by a nonlinear operator [3.4]. It is defined over an a · a rectangle with $a = 2^q$, $q=1,\dots,q_o$ by

$$f_x = \prod_{q=1}^{q_o} f_x^{(2^q)} \tag{11}$$

$$f_x^{(a)} = \frac{1}{a(2a+1)} \left| \sum_{\mu=-a+1}^{o} \sum_{\nu=-a}^{a} f_{i+\mu,j+\nu} - \sum_{\mu=1}^{a} \sum_{\nu=-a}^{a} f_{i+\mu,j+\nu} \right| .$$

In an analogous manner f_y is defined and h_{ij} may be taken as in equ.(9) or as in

$$h_{ij} = \max \{f_x, f_y\} .$$

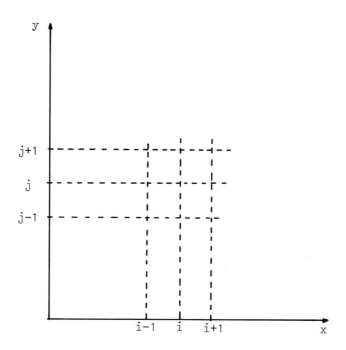

Fig. 2.   Indices of the picture matrix.

Considerable effort was taken to design "optimal" detectors of grey level changes [1.3,3.1]. Two examples are given. One approach is to superimpose a circular window onto the image and to compute the function

$$g(x,y,a_o,a_1,a_2,b,d) = \begin{cases} b & \text{if} \quad a_1 x + a_2 y \le a_o \\ b+d & \text{if} \quad a_1 x + a_2 y > a_o \end{cases} \qquad (12)$$

which best fits to the image inside the circular window. The function g is an ideal edge of arbitrary orientation and stepsize; these are controlled by the parameters a_o, a_1, a_2, b, d. A computationally efficient method is derived in [3.5] to compute these parameters such that the mean square error between the image f and the edge g is minimal over the circular window. Certainly, besides steps there are other changes of grey level like ramps or roofs.

Another approach is to determine a Wiener edge detector which is the least squares filter to implement the Laplacian operator. This is defined as

$$h(x,y) = \frac{\delta^2 f(x,y)}{\delta x^2} + \frac{\delta^2 f(x,y)}{\delta y^2} . \qquad (13)$$

A digital version is

$$h_{ij} = |4f_{ij} - (f_{i,j-1} + f_{i,j+1} + f_{i+1,j} + f_{i-1,j})| .$$

If s is an ideal image with Laplacian h and the observed image is

$$f(x,y) = s(x,y) + n(x,y), \qquad (14)$$

where n is a noise process, a linear filter is derived which yields an estimate $\hat{h}$ of the ideal h. Let

$$G_d(\xi,\eta) = (\xi^2 + \eta^2) \exp(-\xi^2 - \eta^2) \qquad (15)$$

be the transfer function of the desired filter which acts on s(x,y). This G_d is a Laplacian weighted by a low pass filter. Let S_{ss} and S_{nn} be the spectral density functions of the ideal image s and noise n. In this case the optimal filter G_o, acting on f(x,y) and yielding the minimum mean square estimate $\hat{h}$ of h is

$$G_o = \frac{H_d S_{ss}}{S_{ss}+S_{nn}} . \qquad (16)$$

A model of S_{ss} and a recursive filter realization of G_o are given in [3.6]. For a further approach, which fits a plane to f(x,y) and obtains information about grey level changes from the plane parameters, the reader is referred to [1.4,3.7].

The result of the above processing is a matrix h, the elements h_{ij} of which yield a measure of grey level change. If this change exceeds a threshold Θ,

$$h_{ij} \geq 0 \tag{17}$$

then h_{ij} is considered to be a candidate for a point on a contour. Some methods to fit a contour line or curve to these points are discussed next. It should be noted that in addition to equ.(17) information about the direction of the contour is available.

Determination of contours

An approach to finding straight lines is given in [3.3]. In the following a "point" is meant to meet equ.(17). It is tried to fit a line

$$ax + by = c \quad , \quad a^2 + b^2 > o \tag{18}$$

to a number N of points (x_i, y_i). The parameters a, b, c are determined such that the mean square error

$$e = \frac{1}{N} \sum_{i=1}^{N} (ax_i + by_i - c)^2 \tag{19}$$

is minimal. This error is related to vertical distance d of (x_i, y_i) to the line by $d = e/b^2$; it is not the perpendicular distance to the line. Therefore, it is useful to distinguish whether the line is closer to a horizontal or to a vertical line. Abbreviated we write

$$\bar{u} = \sum_{i=1}^{N} u_i \quad ,$$

$$p = \bar{y}^2 - \bar{x}^2 + N(\overline{x^2} - \overline{y^2}) \quad .$$

The parameters are [1.4]

$$\left.\begin{array}{l} a = N \, \overline{xy} - \bar{x} \, \bar{y} \\ b = \bar{x}^2 - N\overline{x^2} \\ c = \frac{1}{N} (a\bar{x} + b\bar{y}) \end{array}\right\} \quad p \geq o \quad , \tag{20}$$

$$\left.\begin{array}{l} a = \bar{y}^2 - N\overline{y^2} \\ b = N \, \overline{xy} - \bar{x} \, \bar{y} \\ c = \frac{1}{N}(a\bar{x} + b\bar{y}) \end{array}\right\} \quad p < o \quad . \tag{21}$$

Starting with two points, parameters a, b, c and sums $\bar{x}$, $\bar{y}$, and so on, are computed. If the distance between the line and a new point is smaller than a threshold, this new point is used to compute an updated line; otherwise a new line is introduced. Updating is easily done, since $\bar{x}$, $\bar{y}$, and so on, are easy to alter. A low threshold will yield a large number of fairly short line segments. In the next step it is tried to join two adjacent segments to one longer line. Again, this is easily done by just adding the respective quantities $\bar{x}$, $\bar{y}$, and so on, obtaining new parameters. Joining is considered successful if the error

$$
s = \begin{cases} \dfrac{b}{N(a^2+b^2)} \ (b\ \overline{y^2} - c\overline{y} + a\ \overline{xy}) & ,\quad p \geq o \\[4mm] \dfrac{a}{N(a^2+b^2)} \ (a\ \overline{x^2} - c\overline{x} + b\ \overline{xy}) & ,\quad p < o \end{cases} \tag{22}
$$

of the longer line does not exceed a threshold. The above error is
the mean squared perpendicular distance of the points from the line.
This process is continued until no longer lines are obtained.

The above method searches through all points which might result
from a contour. If the number of parameters of the curve is low (say,
two or three parameters), it is more efficient to search for the parameter space. This is possible by using the Hough transform [3.8].
The principle is illustrated by a straight line, but it is applicable
to any curve allowing a parametric representation. A straight line
allows, among others, the normal representation

$$
x \cos \alpha + y \sin \alpha = r \ , \tag{23}
$$

with parameters $\alpha(o \leq \alpha \leq \pi)$ and r $(r \geq o)$. A straight line in the
x,y-plane corresponds to one point in the α,r-plane. Alternatively,
one point (x_i,y_i) in the x,y-plane may be viewed as defining a curve
in the α,r-plane. Several points in the x,y-plane are on the same
line if the corresponding curves in the α,r-plane intersect in one
point. This point gives the parameters of the line. To obtain lines
from a set of points in the x,y-plane one proceeds as follows. The
α,r-plane is suitably quantized, for instance, with p · q elements.
It is advantageous that the ranges of α and r are known (certainly,
r will not exceed the size of the image, since r is the perpendicular
distance of the line from the origin). A counter is attributed to any
point α_j,r_k of the quantized α,r-plane. For any point (x_i,y_i) compute α_j, j = 1,...,p and the corresponding $r_k = x_i \cos \alpha_j + y_i \sin \alpha_j$
and increase the counter of α_j,r_k by one. Having done this for all
points (x_i,y_i), the value of the counter of parameter α_j,r_k gives
the number of points on a line with these parameter values. This
principle has been modified, for instance, to detect tumors [3.9]
and ribs [1.10] in chest radiographs.

Short line segments which may lie on a longer straight line or a
curve can be detected by masks or templates. There are various approaches, using different masks, acting on the original picture or
a preprocessed version. An example is [3.10]. Recently, several
methods for iterative enhancement of edge elements were proposed
[3.11 - 3.13]. Only one approach is discussed here [3.12]. The original image f is transformed to an image h containing magnitude p
and orientation q, which are quantized to integer values to yield
an appropriate resolution. Magnitude and orientation may be obtained
by one of the methods discussed above or by other methods [3.14]. A
point h_{ij} of h with magnitude p and orientation q is superimposed by

a mask of n points, such that the n points have a certain orientation. The point h_{ij} will be inspected by the mask corresponding to its orientation q. The magnitudes and orientations of points of h within the mask are p_i, q_i, i=1,...,n. A parameter

$$\rho = \sum_{i=1}^{n} w_{|q_i - q|} \cdot p_i \qquad (24)$$

is computed. It means that the intensities p_i of points inside the mask are weighted and summed. If the weights are positive for small $q_i - q$ and negative for large $q_i - q$, this amounts to large ρ for points with similar orientations within the mask. Therefore, the magnitude p of the point under consideration is increased if $\rho > o$, and decreased if $\rho < o$. This process is iterated. Orientations are also adjusted in an independent algorithm. It is based on the principle that orientations q_i of points within the mask corresponding to the orientation q of the point under consideration should not differ too much from q. If too many points in the mask have significantly different orientations, the existence of a line at h_{ij} is questionable. If many points in the mask have slightly different orientation, the value q is slightly changed.

The above methods for detection of grey level changes and contours are in principal of general applicability. Of course, one method may be more suited to a particular field of problems Ω than another. They are strictly data driven or bottom up and do not make use of any structural knowledge which usually is available. Approaches to the use of such knowledge are, for instance, [3.15,3.16]. So far, only grey level images or one spectral channel were considered. In multispectral images the above methods may be applied to each channel; a modification of edge detection to color images is given in [3.17]. Finally it is mentioned that there are various approaches to efficiently code or characterize contours and boundary lines. Among these are chain code [3.18,3.19], Fourier descriptors [3.20-3.22], and moment invariants [3.23,3.24].

3.2 Regions

Region extraction is in a sense dual to contour extraction because the fundamental aspect of a region is homogeneity with respect to an appropriate criterion. Surveys are given in [3.25,3.26]. Difficulties imposed by noise are analogous to those mentioned in Subsection 3.1. A further difficulty is to find a criterion of homogeneity which is meaningful for the particular problem. Extracting regions from an image or segmentation of an image into regions means that the image array $\underline{f}$ is partitioned into M connected arrays $\underline{f}_\nu$

$$\underline{f} \longrightarrow \{\underline{f}_1, \ldots, \underline{f}_M\} \quad , \qquad (25)$$

such that

$$\bigcup_{i=1}^{M} \underline{f}_i = \underline{f} \quad \text{and} \quad \underline{f}_i \cap \underline{f}_j = \emptyset \quad \text{for} \quad i \neq j.$$

If the criterion of homogeneity is denoted by H and $H(\underline{f}_i)$ is a predicate which is one if $\underline{f}_i$ satisfies H and zero otherwise, it is required that

$$H(\underline{f}_i) = 1 \qquad \text{for } i = 1,\ldots,M \tag{26}$$

$$H(\underline{f}_i \cup \underline{f}_j) = o \quad \text{for } i \neq j \text{ if } \underline{f}_i, \underline{f}_j \text{ are adjacent.}$$

The last requirement states that neighbouring regions must have different properties.

A simple and in some cases sufficient method is to threshold an image at appropriate levels. This is particularly useful if there are objects of similar grey level on some background. In this case the grey level histogram will show two modes, one belonging to the objects, the other to the background. The threshold θ is adjusted to the valley between the modes. Setting

$$h_{ij} = \begin{cases} 1 & \text{if} \quad f_{ij} \geq \theta \\ o & \text{if} \quad f_{ij} < \theta \end{cases} \tag{27}$$

an array $\underline{h}$ is obtained in which objects are marked by 1. This was used, for instance, in chromosome analysis [3.27] and character classification [3.28]. A constant threshold will not be sufficient in many applications. In these cases the threshold may be chosen according to average grey level in a small neighbourhood of f_{ij} or be based on a grey level histogram of only part of the image. A more detailed discussion is given in [1.6].

A more complex criterion than just the grey level of one image point f_{ij} is used in [3.29]. First, the whole image is segmented into cells of size 2 · 2 or 4 · 4 and so on. A statistic of the grey levels of cells is computed next; an example of such a statistic is the histogram. The statistic of the first cell is compared to a neighbouring cell. If the statistics are not similar, the cell is labeled as dissimilar; if the statistics are similar the two cells are merged to one new cell with a newly computed statistic. It is tried to extend the new cell further by examining all of its neighbouring cells and proceeding as indicated. If no more cells can be merged, the new cell is labeled as finished. Then the next unlabeled cell is treated in the same way and so on until all cells are labeled. In this way regions are obtained which are homogeneous according to the grey level statistic. Similarity of statistics may be measured by a statistical test.

A criterion which incorporates the border of two regions to be merged is developed in [3.10]. At first, the image is partitioned into atomic regions, which contain only picture points of equal grey

level. These atomic regions will be fairly small because due to noise
and other imperfections even a "homogeneous" surface will contain different grey levels. If two neighboring picture points belong to different regions, a boundary segment is introduced between them. The
strength of a boundary segment is defined as the difference of the
grey levels of the two neighbouring points. The boundary segment is
weak if its strength does not exceed a threshold θ. The length l
of the weak part of the boundary between two regions is the number
of its weak boundary segments. Two adjacent regions $\underline{f}_1, \underline{f}_2$ with peri-
meters l_1, l_2 are merged if

$$l/l_{min} > \theta_1$$

$$l_{min} = min\{l_1, l_2\} \quad . \tag{28}$$

This requirement, together with the choice of θ_1, allows to control the length l_r of the boundary resulting from merging regions
$\underline{f}_1, \underline{f}_2$. If $\theta_1 > 0,5$ the resulting boundary must be smaller than max
(l_1, l_2); if $\theta_1 < 0,5$ it may be larger. A second heuristic is to merge
two adjacent regions with the common boundary of length l_c if

$$l/l_c > \theta_2 \quad . \tag{29}$$

The first heuristic, see equ. (28), evaluates more global information than the second which is more local. To obtain satisfactory results it is, therefore, not possible to use only the simpler second
heuristic.

The above two methods start with a large number of small regions
which are merged to larger ones. Another possibility is to start with
the entire image and to split it into homogeneous regions. This is done
in [3.30]. The first region is the whole image. A region is homogeneous
if the mean grey level of any subregion is equal to the mean grey level of the region. As long as there are inhomogeneous regions these
are subdivided to get more homogeneous regions. It is shown in [3.30]
that in order to test a region for homogeneity it is sufficient to
test two arbitrary subregions for equal mean grey value. The application to M-channel multispectral images is straightforward if the
mean grey level is replaced by a vector of M mean values. To obtain
two subregions a criterion of partition efficiency is introduced and
several partitions are tried to find a good one.

It is also possible to use functional approximation to find regions [3.31] and combine split and merge processes [3.32]. The above
examples on region extraction are strictly data driven or bottom up;
no structural information about the underlying images is used. Approaches to the use of such information are given in [3.33,3.34].
Since no example was provided for the use of structural information
in contour detection it seems appropriate to give such an example
for region extraction.

The basis of the approach in [3.33] is to find regions such that the probability of correct description of the images is maximized; this probability is influenced by a priori information (knowledge) about the images and measurements on a particular image. This results, for an image $^\rho\underline{f}$, in the requirement

$$P(^\rho B | \text{information, measurements}) = \max \quad . \tag{30}$$

If we have M regions $\underline{f}_i$ as in equ. (25) these will be attached meanings Ω^i in the description $^\rho B$, where each Ω^i may be one out of k labels or class names, i.e., $\Omega^i \in \{\Omega_\kappa | \kappa=1,\ldots,k\}$; this is abbreviated by $\underline{f}_i \longrightarrow \Omega^i$. Examples are the labels "water", "tree", or "sky" for regions. The boundary between neighbouring regions is denoted by $b(i,j)$. It contains features like direction and smoothness. Under appropriate independence assumptions equ. (30) is simplified to

$$\prod_{i=1}^{M} P(\underline{f}_i \longrightarrow \Omega^i | \text{measurements on } \underline{f}_i) \quad , \tag{31}$$

$$\prod_{i,j} P(b(i,j) \text{ is between } \underline{f}_i \text{ and } \underline{f}_j | \text{measurements on } b(i,j)) = \max \quad .$$

The process starts with many small regions, which are preliminarily merged similar to [3.10]. Regions are merged in a way that equ. (31) is increased. The main effort in [3.33] was to develop an efficient algorithm to obtain a nearly optimal set of regions and to estimate the various probabilities. In this approach merging of regions and description (or interpretation) of the image are combined to influence each other.

3.3   Textures

Texture is a property of a surface of an object and, therefore, is related to the last subsection. However, because it is more complex than just grey level or color and not yet well understood, it is treated seperately. Pictorial examples of textures are given in [3.35]. Some aspects are discussed in the survey of [3.26]. Interesting aspects of human texture discrimination are investigated in [3.36]. Intuitively, texture is characterized by random or deterministic distribution of a textural primitive. The problem is to determine such a primitive; this is analogous to the problem of finding features in a pattern classification task. The size of textural primitives usually is not known in advance, so different attempts with different sizes may be necessary. Textural information is used in the classification of regions as well as in region extraction. An extensive evaluation of different texture measures is given in [3.37] and a review in [3.38].

The intuitive idea of repetition of textural primitives is the basis for defining $L \cdot L$ spatial-dependence matrices $P(d,\alpha)$, as in

[3.39]; (L is the number of grey levels). This matrix is

$$P(d,\alpha) = [\![p_{\mu\nu}(d,\alpha)]\!] \quad .$$ (32)

The elements $p_{\mu\nu}(d,\alpha)$ give the number of picture points $f_{ij}=\mu$ and $f_{kl}=\nu$, where f_{ij} and f_{kl} have distance d and orientation α. It is easy to compute this matrix for an image f. The matrices may be computed for different values of d and α and used to obtain textural features. Four of the fourteen features defined in [3.39] are given as an example. For matrix $P(d,\alpha)$, define the normalized elements

$$P_{\mu\nu} = \frac{p_{\mu\nu}(d,\alpha)}{\displaystyle\sum_{\mu} \sum_{\nu} p_{\mu\nu}(d,\alpha)} \quad .$$ (33)

Then the features are

$$c_1 = \sum_{\mu=0}^{L-1} \sum_{\nu=0}^{L-1} P_{\mu\nu}^{\;2} \quad,$$

$$c_2 = \sum_{l=0}^{L-1} l^2 \sum_{|\mu-\nu|=1} P_{\mu\nu} \quad,$$

$$c_3 = \left[\sum_{\mu} \sum_{\nu} \mu\nu P_{\mu\nu} - m_\mu m_\nu \right] \Big/ \sigma_\mu \sigma_\nu \quad,$$ (34)

$$c_4 = - \sum_{\mu} \sum_{\nu} P_{\mu\nu} \log P_{\mu\nu} \quad .$$

In the above equation m_μ and σ_μ are mean and standard deviation of the marginal distribution

$$p_\nu^{(\mu)} = \sum_{\mu=0}^{L-1} P_{\mu\nu} \quad,$$ (35)

and m_ν, σ_ν are defined analogously. The four features are termed angular second moment, contrast, correlation, and entropy.

Related work is found in [3.40,3.41]. The idea to compute statistical parameters of the images may be varied. For instance, instead of taking grey levels μ and ν, as in equ.(32), only the difference $l=|\mu-\nu|$ may be considered [3.37]. The use of grey level run length is studied in [3.42]. The grey level run-length matrix

$$G(\alpha) = [\![g_{\mu l}(\alpha)]\!]$$ (36)

contains the number $g_{\mu l}(\alpha)$ of run length l at grey level μ in direction α. The run length l is the number of collinear picture points with constant grey level; these may be determined in different directions α.

With

$$N_t = \sum_\mu \sum_l g_{\mu l} \quad , \tag{37}$$

five textural features are defined. They are short-run emphasis, long run emphasis, grey level nonuniformity, run-length nonuniformity and run percentage, and are given by

$$c_1 = \sum_\mu \sum_l \frac{g_{\mu l}}{l^2} \Big/ N_t$$

$$c_2 = \sum_\mu \sum_l l^2 \, g_{\mu l} \Big/ N_t$$

$$c_3 = \sum_\mu \left(\sum_l g_{\mu l} \right)^2 \Big/ N_t \tag{38}$$

$$c_4 = \sum_l \left(\sum_\mu g_{\mu l} \right)^2 \Big/ N_t$$

$$c_5 = \sum_\mu \sum_l g_{\mu l} \Big/ N_p \quad .$$

N_p is the number of possible runs if all had length one.

Texture discrimination by means of histograms of local properties is done in [3.43,3.44]. The property used in [3.44] is a spot detector, giving

$$h_{ij} = \left| \frac{1}{2(N+1)} \sum_{\mu=i-N}^{i+N} \sum_{\nu=j-N}^{j+N} f_{\mu\nu} - \frac{1}{2(M+1)} \sum_{\mu=i-M}^{i+M} \sum_{\nu=j-M}^{j+M} f_{\mu\nu} \right| . \tag{39}$$

It is the difference in average grey value of two square regions of size $(2N+1)^2$ and $(2M+1)^2$, $M > N$, which are centered at (i,j). A value h_{ij} is suppressed if there is inside the square of size $(2M+1)^2$ another h_{kl}^{ij} with $h_{kl} > h_{ij}$. This nonmaximum suppression is used to obtain bimodal histograms with the images used. The two modes correspond to two different textures which thus may be discriminated. The parameters M and N are chosen so as to yield strongest bimodality.

Also, other texture measures were developed, like spectral features [3.37], intensity profiles [3.45], and relative frequency of local extrema in intensity [3.46]. The spectral features are obtained from $|F|^2$, where $F(\xi,\eta)$ is the Fourier transform of image $f(x,y)$. They are averages over ring or wedge-shaped areas centered in the origin of the (ξ,η)-plane [3.47,3.48].

Application of the above methods to texture discrimination presents no problem if an image of uniform texture (for instance, containing only water, swamp, or woods) is provided. However, if an image contains several areas of different texture, as

usually will be the case, the image first has to be segmented into regions of homogeneous texture. In this case it would not be meaningful to compute spatial dependence matrices or spectral features or
something else for the entire image. The textural features may be
obtained for small areas of the image and used to decide to which
type of texture they belong.

4. CLASSIFICATION

There are several important applications where interesting parts
of an image can be isolated and classified by standard techniques.
Among these are image analysis in cytology and remote sensing. As
mentioned, these applications are treated in detail in [1.5]. Therefore, a short account of numerical classification methods is given
in the following. An early survey is given in [4.1], general text-
books are [1.3,1.4,4.2-4.4].

If an object, a region of an image or, in general, a pattern is
to be classified, an n-dimensional feature-vector $\underline{c}$ is extracted from
the pattern and mapped to one out of k integers, i.e.,

$$\underline{c} \longrightarrow \kappa \in \{1,\ldots,k\} \quad , \tag{40}$$

where integer κ denotes pattern class Ω_κ. In the case of M-channel
multispectral images each picture point (consisting of M measurements
in the respective channels) may be considered as a feature vector
with M=n and classified individually [4.5]. Some other methods to
find feature vectors were discussed in Section 3, for instance, Fourier
descriptors of boundaries or textural features. Another standard approach to localization of an interesting object is template matching.
It means that a prototype or template of the object is moved across
the image to see whether there is a good fit between object and a sub-
area of the image. This process can be implemented in the space or
frequency domain. To overcome computational difficulties hierarchical
matching processes were employed [4.6-4.9]. However, we shall limit
the discussion to standard classification methods.

A classification in the sense of equ.(40) can be accomplished
by finding k functions (decision functions, discrimination functions)
$d(\underline{c},\underline{a}_\kappa)$, $\kappa=1,\ldots,k$. The classifier works according to the decision
rule

$$d(\underline{c},\underline{a}_\kappa) = \max_\lambda d(\underline{c},\underline{a}_\lambda) \longrightarrow \underline{c} \in \Omega_\kappa \quad . \tag{41}$$

The functions d depend on parameter vectors $\underline{a}_\kappa$; these are adjusted
in order to make decisions as reliable as possible.

A reliable decision may, among others, be defined as the decision yielding minimal probability of error p_e. If the a priori
probability of class Ω_κ is p_κ and the class-conditional density of

$\underline{c}$ is $w(\underline{c}|\Omega_\kappa)$, then a special version of equ. (41), which minimizes p_e, is

$$p_\kappa w(\underline{c}|\Omega_\kappa) = \max_\lambda p_\lambda w(\underline{c}|\Omega_\lambda) \longrightarrow \underline{c} \in \Omega_\kappa \quad . \tag{42}$$

The problem now is to determine the k densities $w(\underline{c}|\Omega_\kappa)$. It seems that there are only two feasible approaches to this problem. The first is to assume that the one-dimensional densities $w(c_\nu|\Omega_\kappa)$ of the features $c_\nu, \nu = 1,\ldots,n$ are independent. In this case,

$$w(\underline{c}|\Omega_\kappa) = \prod_{\nu=1}^{n} w(c_\nu|\Omega_\kappa) \quad , \tag{43}$$

and $w(c_\nu|\Omega_\kappa)$ may be approximated by a histogram or a series expansion. The second approach is to assume that the feature vector $\underline{c}$ has an n-dimensional Gaussian density. In this case equ. (41) becomes

$$d(\underline{c},\underline{a}_\kappa) = d(\underline{c},\underline{\mu}_\kappa,\underline{K}_\kappa)$$

$$= -\frac{1}{2} (\underline{c}-\underline{\mu}_\kappa)_t \underline{K}_\kappa^{-1} (\underline{c}-\underline{\mu}_\kappa) + \ln \frac{p_\kappa}{\sqrt{|2\pi\underline{K}_\kappa|}} \quad . \tag{44}$$

The unknown parameters $\underline{\mu}_\kappa, \underline{K}_\kappa$ are estimated from a classified sample of patterns.

If information about probability densities is not available, it is reasonable to try a direct approximation of $d(\underline{c},\underline{a}_\kappa)$. Let

$$\underline{d}(\underline{c}) = (d_1(\underline{c}),\ldots,d_k(\underline{c}))_t \tag{45}$$

be ideal decision functions defined by

$$d_\kappa(\underline{c}) = \begin{cases} 1 & \text{if} \quad \underline{c} \in \Omega_\kappa \\ o & \text{otherwise} \end{cases} \quad , \quad \kappa = 1,\ldots,k \quad . \tag{46}$$

This ideal function is approximated by

$$\underline{d}'(\underline{c}) = \underline{A}_t \underline{\phi}(\underline{c}) \quad , \tag{47}$$

where $\underline{A}$ is a matrix of coefficients (the columns of $\underline{A}$ are the parameter vectors $\underline{a}_\kappa$ from equ. (41) and $\underline{\phi}$ is a vector of m independent functions $\phi_\mu(\underline{c})$, $\mu = 1,\ldots,m$. If it is tried to approximate $\underline{d}$ with least mean-square error, matrix $\underline{A}$ has to be chosen in order to minimize

$$E\{(\underline{d} - \underline{A}_t\underline{\phi}(\underline{c}))_t (\underline{d} - \underline{A}_t\underline{\phi}(\underline{c}))\} = \min_{\underline{A}} \; ; \tag{48}$$

$E\{\ldots\}$ denotes expected value. This minimization is done by choosing

$$\underline{A} = \left[E\{\underline{\phi}(\underline{c})\underline{\phi}_t(\underline{c})\}\right]^{-1} E\{\underline{\phi}(\underline{c})\underline{d}_t(\underline{c})\} \quad . \tag{49}$$

Again, the expectations in equ. (49) can be estimated from a classified sample. By proper choice of ϕ_μ, one obtains linear, quadratic or other classifiers.

There are some classifiers which do not quite fit to equ. (41),
as for instance, nonparametric and hierarchical classifiers [4.10,4.11].
More important, however, is that there are cases in which classified
samples are not available; it is also possible that it is desirable
to find classes which are inherent to a sample of unclassified patterns. In this case unsupervised classification methods are used. It
is not possible to discuss these techniques in any detail here. Although it is theoretically possible to identify the components of a
finite mixture of normal distributions, this process presently is not
computationally feasible. Therefore, decision-directed methods are
applied [1.3,1.4] or methods of cluster analysis; the latter have
been used intensively in the analysis of multispectral images [4.12].
Algorithms used are of the ISODATA-type [4.13].

5. FINAL ANALYSIS

So far, discussion of image analysis has proceeded to the point
where boundaries, regions, and textured areas may be extracted. Also,
some of the simpler constituents may be classified by numerical methods.
Final analysis now has to attach meaning to these simple constituents
in such a manner that a globally correct description ^{p}B of image $^{p}\underline{f}$ results. Not only should an isolated object be labeled with a certain
meaning independent of surrounding constituents but should also be done
with respect to relevant context of the image and all available information about the field of problems. This is what makes analysis extremely difficult. On the other hand, taking account of such information will allow resolving ambiguities and to correct errors which
usually will result during extraction of simpler constituents and/or
independent classification [5.1].

Discussion of analysis methods is started with a general framework followed by special implementations. This general start seems
appropriate in order to get a formal view of the problem of analysis,
which is independent of detailed problems inherent in particular
methods; of the latter, syntactical methods, graph matching, and relaxation labeling will be treated.

5.1 General Framework

In Fig.1 a diagram of pictorial data flow in an image-analysis system
was presented. It was pointed out that this was not necessarily the
direction of flow of control; this point is emphasized again since
the presentation of the paper follows Fig.1. A more flexible system,

allowing fairly arbitrary flow of data and control is shown in
Fig. 3. It consists of four main parts:

1. A data base which contains <u>results</u> about ρ_f which were obtained
   so far. This comprises spatial information (e.g.,where is a contour
   line), possible alternatives (e.g. there may be a river or a highway), and different levels of representation (e.g., an array of
   picture points at one level, two parallel lines at another level,
   a bridge at the highest level).

2. A module which executes <u>control</u>; it decides which methods to apply,
   which information to use and how to coordinate access of the
   results data base.

3. A data base which contains <u>methods</u> for preprocessing an image, extraction of simple constituents and classification.

4. A data base which contains <u>information</u> (knowledge) about structural
   properties of images and possibly about the field of problems.

It is expected that the outline of Fig.3 will help to distinguish
methods, information and control in an image analysis system. There
are similarities to the approaches in [5.2,5.3] and the results in
[5.4]. Often these parts of an analysis system are combined to one
program. This may be the most efficient way (in the sense of computing
time) to solve a particular problem, whereas for research purposes,
flexible system structure, error handling and program maintenance
a clear distinction will be advantageous. As indicated in Fig. 3, for
instance, the information data base may be structured into several
blocks. If these blocks are independent of each other (e.g.,infor-
mation about contour lines built up by image points, regions, and objects built up by contour lines), the control module may invoke these
blocks in parallel. This allows parallel execution of several operations on the image; of course, careful control of access to the results
data base by the parallel processes is necessary. An example of this
kind of organization, although not of image analysis, is given in
[5.5]; an approach to image analysis in [5.37].

Presently, information about structural properties of images
is usually gathered by the designer of the system, who looks at sample
pictures. Information is represented in rules of a formal grammar (see
Subsection 5.2), in form of a graph (see Subsection 5.3), or in compatibilities of relaxation methods (see Subsection 5.4). In the case of
formal grammars, methods of automatic inference of grammars have been
developed which will be mentioned later. A fairly general representation
is provided by production rules which have a general IF - THEN format [5.2,5.6]. The IF term contains an arbitrary premise or logical
statement, the THEN term one or more actions or conclusions. An example
would be:

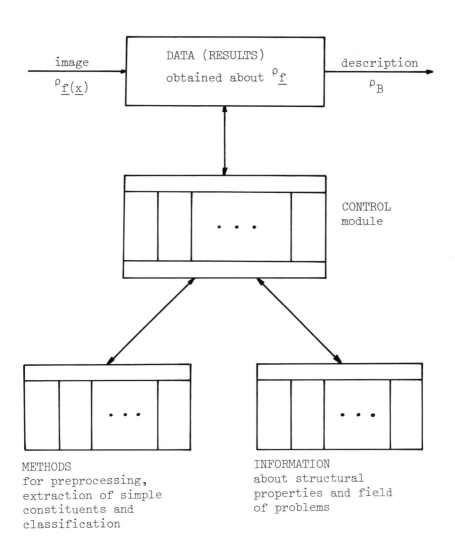

Fig. 3.   A system for image analysis.

> IF   :   there are two long, closely adjacent, nearly parallel
>   lines; these are crossed at nearly right angle by two
>   other long, closely adjacent, nearly parallel lines.
>
> THEN :   there is evidence that these may be a crossing of (50)
>   two highways; a highway crossing a river; or a rail-
>   road crossing a river.

Clearly, a rule

$$\beta_\nu \longrightarrow \gamma_\nu \qquad\qquad\qquad\qquad (51)$$

of a formal grammar is of the same form, although more specialized.
It means that in a string of symbols substring β_ν may be replaced by
γ_ν. Also, there is a strong resemblance between representation of information in graphs (or semantic nets) and production rules. The graph
of Fig.4 may be expressed by the rule

> If   :   there are green leaves above a trunk,
>
> THEN :   this is a tree.

A straightforward generalization of equ.(50) is to include an
advice about the actions or production rules to be examined next.
This may be done by taking rules one after another or as in programmed
grammars. This results in a production system which embodies a certain
control structure [5.6].

Let TRUE denote the case that the IF term of a production rule
is satisfied and FALSE that it is not. A general production rule is

> IF   : (premise or logical statement),
> THEN   : (action or conclusion),
> TRUE   : (next rule or action), (52)
> FALSE : (next rule or action).

Apparently control and representation of information are closely
related. Production rules of the form of equ.(52) may be used to represent information and to realize a certain control structure. In
this case, both are intimately related. On the other hand, if production rules contain no information about the order of their application, this has to be done by the control module. A good compromise
might be as indicated in Fig.3. The control module activates one or
several blocks of information data base. These blocks execute their
processes and return results to the results data base. After a process has finished the control module decides about the next block of
actions to be taken.

Another general approach to representation of the control module
is introduced by abstract programs [5.8] and flowchart schemes [5.9].
An application to analysis of radiographs are the "image analysis
graphs" [5.10]. It is stated in [5.9] that "essentially, a flowchart

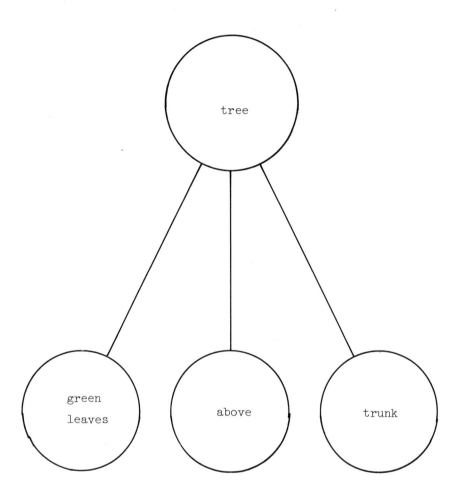

Fig. 4. Representation of information in a graph.

scheme depicts the control structure of the program, leaving much of
the details to be specified in an interpretation". Such a scheme is
represented by a graph or a flowchart. It is made up of an alphabet
and statements. The alphabet consists of function constants g_i, pre-
dicate constants p_i, input variables $\underline{f}$, program variables $\underline{y}$ (for intermediate storage), and output variables B. In the abstract form the
particular mapping provided by g_i, p_i is left unspecified. The statements are

1. START   : $\underline{y} \leftarrow g_o(\underline{f})$,
2. ASSIGN : $\underline{y} \leftarrow g_i(\underline{f},\underline{y})$,   (53)
3. TEST   : IF $p_i(\underline{x},\underline{y})$ TRUE GO TO LABEL i, ELSE GO TO j,
4. HALT   : $B \leftarrow g(\underline{f},\underline{y})$.

The representation of these statements by blocks of a flowchart is
obvious and may be omitted here. A simplified example is the abstract
program

1. READ IMAGE f : $\underline{y}_1 \leftarrow \underline{f}$,
2. PREPROCESS f : $\underline{y}_2 \leftarrow g_1(\underline{y}_1)$,
3. EXTRACT BOUNDARIES of f : $\underline{y}_3 \leftarrow g_2(\underline{y}_2)$,   (54)
4. ANALYZE f : $\underline{y}_4 \leftarrow g_3(\underline{y}_3)$,
5. SUCCESS: IF (no error)   GO TO 6, ELSE FAILURE ,
6. WRITE DESCRIPTION $B \leftarrow \underline{y}_4$.

It is apparent that Fig. 1 thus may be viewed as a particular
control structure, passing control from one block to the other in a
fixed manner. Representation of control structure in a flowchart
scheme is particularly useful for more involved examples. It might
be desirable to develop a more sophisticated control structure which
gives control back to "boundary extraction" or "preprocessing" if
"analysis" fails at some step; for instance, preprocessing could be
tried again with a modified filter for noise removal.

A flowchart scheme is converted to a program if an interpretation
of the scheme is provided. This requires specification of a domain
for the variables, function mappings for g_i, and predicate mappings
for p_i. An example is:

abstract form : $\longrightarrow$ interpreted form :

PREPROCESS $\underline{f}$: $h_{ij} = 10 \log f_{ij}$ (55)

$\underline{y}_2 = \underline{h}$

In the above case, the abstract command "replace $\underline{f}$ by a preprocessed image $\underline{y}_2$" is interpreted as the logarithmic scaling of
picture points.

Finally, to carry out an actual computation, it is necessary that

values of the input variables are given. In our case an image array
has to be provided.

Introduction of flowchart schemes should help to distinguish
certain aspects of analysis as depicted in Fig. 3. Of course, it does
not give an algorithm to automatically design control structures. So
far this design is left to the scientist engaged in image analysis.
It seems appropriate, therefore, to give some specific examples in
the next three subsections. Although intensive research efforts and
remarkable achievements have been made in image analysis this area
is still wide open for further work.

5.2 Syntactical Methods

Application of formal languages to pattern classification and
analysis has proved to be useful in different applications. A language
L is generated by a formal grammar

$$G = (V_T, V_N, R, S) \quad , \tag{56}$$

where V_T, V_N are disjoint finite sets of terminal and nonterminal
symbols, R is a finite set of productions or rules

$$r_i : \beta_i \longrightarrow \gamma_i \quad ; \quad \beta_i \in (V_N \cup V_T)^+ \quad ; \quad \gamma_i \in (V_N \cup V_T)^*, \tag{57}$$

and $S \in V_N$ is the starting symbol. In addition, β_i is required to
contain at least one nonterminal symbol. Starting with S, rules
$r_i \in R$ are applied in arbitrary order to obtain a string $v \in V_T^*$ of
terminal symbols; this is denoted by $S \stackrel{*}{\Longrightarrow} v$. The language $L(G)$
generated by G is the set of terminal strings (or sentences) which
can be obtained from S by application of rules; this language is

$$L(G) = \{v \mid S \stackrel{*}{\Longrightarrow} v, v \in V_T^*\} \quad . \tag{58}$$

The theory of formal languages is well developed and may be
obtained from References [5.11,5.12].

When applied to image analysis the terminal symbols are simpler
constituents of the image and their respective relations. An example
would be:

$$V_T = \{s_1, s_2, s_3, s_4, s_5\}$$

s_1 : horizontal line $\left.\right)$
s_2 : vertical line $\left.\right\}$ simple constituents   (59)
s_3 : circle $\left.\right)$
s_4 : above $\left.\right\}$ relations
s_5 : adjacent to

It is an open problem to systematically generate those terminals which
are best suited to a particular field of problems. It is important to

note that the grammar of equ. (56) is a finite scheme to represent a
possibly infinite set of sentences (in our case images) which obey
certain rules; to put it another way, the images generated by G have
a certain structure. It is thus possible to exploit assumption 3 of
Section 1. Another important point is that generation of a sentence
v may be reversed; given a sentence v it can be parsed to yield the
rules which generated v, or v is rejected as not belonging to L(G).
These methods, too, are well known [5.13,5.14]. There are numerous
extensions and modifications of grammars which were applied to image
analysis, like programmed grammars [5.15], stochastic grammars [5.16],
tree grammars [5.17], augmented transition network grammars [5.18],
array grammars [5.19], web grammars [5.20], and graph grammars [5.21].
It is impossible to present details of these approaches. However, the
basic idea introduced in equ. (56) and (57) is maintained: substitute
a string (or something else like a tree, a web, a graph, ...) upon
its occurrence by some other string (or something else).

It is evident from the above discussion that information about
structural properties of images is contained in the grammar G, particularly in the rules of equ. (57). Some methods of automatic inference
(or learning) of structural information are available; a survey is
given in [5.22]. Also, interactive learning facilities were developed [5.23].

Parsing a string of symbols is a common task of compilers. It is
possible to work bottom-up or top-down. Bottom-up parsing starts with
the sentence v and tries to apply rules in reverse until the start
symbol S is reached. If this is impossible v is not a sentence of
L(G) and is rejected. Top-down parsing starts with start symbol S
and tries to generate v by applying rules. Again, if this is impossible
v is rejected. There are some important points to note in the case of
image analysis. First, image analysis starts with a picture matrix,
not a string of symbols. Symbols have to be extracted from the image.
Second, this extraction may be erroneous. It is extremely important
that analysis of the symbols allows some kind of error correction or
new attempts of symbol extraction. It is seen that the same structural
information (a grammar G) may be used with different control strategies
(top-down or bottom-up).

An early example of bottom-up analysis is provided by chromosome
analysis [5.24]. In this case simple constituents are boundary segments of various shapes, the only relation is concatenation. A string
of boundary segments coded by symbols is presented to the analyser
which works with a context-free grammar represented in tabular form.
It systematically applies productions and alternatives to the string
in order to reduce it to the start symbol. Approaches to error correction are given in [5.16,5.25]. An error in analysis will occur if
either a symbol s_i (simple constituent) in a string v is substituted
by another symbol s_j or if s_i is deleted, or if in string v an additional symbol s_k is inserted. Augmenting the grammar by these

error productions allows to include into the language certain erroneous strings which may result from errors during extraction of simple constituents. Introduction of stochastic grammars allows the assignment of probabilities to the above mentioned three types of errors. Let G be the grammar in the error-free case, and let $v \in L(G)$ be an error-free string. Let $w \notin L(G)$ be an erroneous string. It is possible to derive an algorithm which yields a string $v^* \in L(G)$ such that

$$p(w|G) = p(w|v^*) \, p(v^*) \tag{60}$$
$$= \max_v \{p(w|v) \, p(v)\} \ .$$

This means that the string $v^* \in L(G)$ is obtained which most probably was converted to string $w \notin L(G)$ by errors during extraction of simple constituents. An example of a top-down analysis is provided in [5.26, 5.27]. The parsing algorithm accepts a context-free grammar and tries to parse a picture with this grammar. Control is top-down; first a hypothesis is generated about a simple constituent which should be present within a certain area of the image. Then it is tried to find this in the image. The image is not converted to a string of symbols prior to analysis. Generation of hypothesis allows to search in a limited area of the image for a limited set of symbols. Thus symbol extraction can be planned by use of structural information. If during analysis several symbols have been detected and these symbols strongly suggest the occurrence of another - yet undetected - symbol, this evidence may be used to influence the search procedure for the new symbol (for instance, by adjusting thresholds). This allows avoidance of errors by use of context. Other examples of applications of syntactical methods are found in [1.7,1.12].

5.3 Graph Methods

As indicated in Subsection 5.1, structural information is representable by a graph. A graph G consists of a set of nodes (or vertices) K and a set of edges $E \subseteq K \times K$. In image analysis, nodes may be simple constituents and edges may be relations between constituents. Of course, it is also possible to represent lines in an image by edges of a graph and points of line intersection in the image by nodes of the graph. The important point is to establish a unique correspondence between image $\underline{f}$ and graph G. In order to prove structural identity (not point by point identity) of two images $\underline{f}_1$, $\underline{f}_2$ represented by graphs G_1, G_2 one has to find an isomorphism between G_1 and G_2. Two graphs are isomorphic if there is a one-to-one node mapping between them which preserves a one-to-one correspondence of edges. In order to find an interesting object $\underline{f}_1$ with graph representation G_1 in a more complex image $\underline{f}$ with graph G it is necessary to find a subgraph G' within G such that G' is isomorphic to G_1. Graph G' is a subgraph of graph G if G' can be obtained from G by removing some nodes from G and all edges adjacent to the removed nodes. Although we do not see

fundamental differences between representation of information in graphs
or rules of grammars, methods to use this information are different;
establishing graph isomorphism needs methods different from parsing
a string of symbols. Graph isomorphisms and graphs in image analysis
are treated, for instance, in [5.10,5.28-5.32].

Testing for isomorphism is facilitated by the fact that nodes and
edges of a graph, which represents an image, have a certain meaning.
Let, for instance, G and G' have node sets K and K'. If the node
$k_1 \in K$ is considered, not any node $k' \in K'$ is a candidate for correspondence to k_1. If, for instance, k_1 represents a "red circular region"
of an image, only nodes $k'_\nu \in K'$ which, too, represent red circular
regions may correspond to k_1. A similar argument holds for edges. It
is thus possible to reduce the amount of search necessary to prove
(or disprove) an isomorphism. This idea can be extended to larger
entities. Let there be two nodes $k_1, k_2 \in K$ related by an edge $e_1 \in E$
with, for instance, the meaning that a red circular region (k_1) is
above (e_1) a black square (k_2). Clearly the triple (k_1, k_2, e_1) of G
can correspond only to those triples of G' with the same meaning (or
labeling). A more formal treatment of this outline may be found in
[5.28,5.31]. An interesting modification, the "structural isomorphism"
is introduced in [5.30]; it allows very efficient structure matching.

Extensive use of graph representations is made in [5.10] for information representation as well as control flowcharting. Analysis is
top down directed. A very interesting point is adjustment of parameters for region extraction according to prior success or failure
of analysis. This allows error recovery as well as use of context information.

5.4 Relaxation Methods

Relaxation methods are of fairly broad applicability. They have
already been mentioned for iterative edge enhancement in Subsection
3.1. The aim of relaxation methods is to find a consistent labeling
of simple constituents (or parts, or objects) of an image. Initially
each constituent may have several possible labels. For a particular
pair of constituents not every pair of labels will be meaningful.
Probabilities of labels are iteratively changed in order to increase
probability of meaningful pairs and to decrease probability of not
meaningful pairs. To put it another way, simple constituents are
first classified independently allowing several class names. By taking
account of the context one tries to obtain an unambiguous and reliable classification of the whole set of constituents. A mathematical
treatment of convergence properties of several relaxation methods is
given in [5.33], an overview of several applications in [5.34].

Among the different relaxation schemes developed so far we shall

give a short outline of a discrete version [5.33,5.35] and an introduction to a nonlinear probabilistic version [5.33]. Suppose a set
$V_T = \{s_1,...,s_n\}$ of simple constituents (which may, for instance, be
regions, or contours, or both) was extracted from image $\underline{f}$. Based on
a priori knowledge about the field of problems Ω it is known that the
set of possible labels is $L = \{l_1,...,l_k\}$. Of course, the labels may
be considered as class names (Section 4). The set of labels which is
possible for a particular object $S_i \in V_T$ is $L_i \subseteq L$; it is determined
only from evidence about s_i. For instance, a label "lake" is not
compatible with a region s_i of color "red" and thus this label is
excluded from L_i. For each pair $s_i,s_j,i \neq j$ of constituents some pairs
of labels will be impossible. By $L_{ij} \subseteq L_i \times L_j$ we denote the set of
pairs of labels which is compatible with s_i, s_j. For instance, a
pair "car, sky" would not be possible for a "yellow region surrounded
by a blue region" but the pair "boat, lake" is compatible. A labeling
$\tilde{L} = \{L_1,...,L_n\}$ assigns a set L_i of labels to each constituent s_i.
This assignment is called consistent, if

$$(l_i \times L_j) \cap L_{ij} \neq \emptyset , \qquad i,j = 1,...,n \quad \text{and} \quad l_i \in L_i . \qquad (61)$$

This equation states that every label in L_i must be possible for s_i;
further that for each pair of constituents s_i,s_j and each label
$l_i \in L_i$ there must be a label $l_j \in L_j$ which is compatible with l_i.
To obtain such a consistent labeling one proceeds as follows:

1. Start with an initial labeling $\tilde{L}^0$.

2. To obtain $\tilde{L}^{N+1}$ from $\tilde{L}^N$ remove from each $L_i^N \in \tilde{L}^N$ all the labels
   l_i which violate equ.(61) for some j.

3. Iterate step 2 until no more labels are removed.

Obviously, if no more labels are removed from a labeling $\tilde{L}^N$ then
$\tilde{L}^N$ is consistent and since all sets are finite the algorithm stops
for a finite N. It is not guaranteed that this final labeling will
be unambiguous. By search procedure it is possible to find unambiguous
labelings provided they exist.

The discrete decision of removing or not removing a label is
avoided if labels are assigned weights between zero and one. By the relaxation process these weights are altered continuously according to
compatibility constraints. In addition to the set L_i of labels assigned
to each constituent s_i a set of probabilities $p_i(l_\mu)$ is assigned to
each label $l_\mu \in L_i$, such that

$$0 \leq p_i(l_\mu) \leq 1 \quad ; \quad \sum_\mu p_i(l_\mu) = 1 \qquad (62)$$

Furthermore, pairs of objects s_i,s_j are assigned compatibilities
$r_{ij}(l_\mu,l_\nu)$ beween -1 and $+1$. A negative value indicates that s_i with
label l_μ and s_j with label l_ν seldom cooccur, a positive value indicates that they often co-occur and a value close to zero means that
they are fairly independent. The compatibilities may thus be inter-

preted as correlations; methods to obtain these values are discussed
in [5.36]. A nonlinear relaxation process is defined by the equations

$$\beta_{i,N}(1_\mu) = \sum_j \alpha_{ij} \sum_\nu r_{ij}(1_\mu,1_\nu) \, p_{j,N}(1_\nu) \quad , \tag{63}$$

$$p_{i,N+1}(1_\mu) = \frac{p_{iN}(1_\mu)\,(1+\beta_{i,N}(1_\mu))}{\sum_\mu p_{iN}(1_\mu)\,(1+\beta_{i,N}(1_\mu))} \quad . \tag{64}$$

In the above equations the subscript N denotes iteration step. The
coefficients α_{ij} are required to satisfy $\sum_j \alpha_{ij} = 1$. A high value
of $p_{j,N}(1_\nu)$ together with a high compatibility $r_{ij}(1_\mu,1_\nu)$ in equ.(63)
will result in a high positive contribution to $\beta_{i,N}(1_\mu)$. Together
with equ.(64) this assures that $p_{i,N+1}(1_\mu)$ is increased if other
constituents with highly probable labels are highly compatible
with label 1_μ of object s_i. If other constituents with highly probable
labels are highly incompatible $p_{i,N+1}(1_\mu)$ will be decreased. Other objects with low probability labels will have little influence. This
behaviour of $p_{i,N+1}(1_\mu)$ is desirable. Furthermore it is guaranteed
that the $p_{i,N+1}(1_\mu)$ are probabilities, i.e. are nonnegative and sum
to unity.   Linear relaxation processes are possible, too. They
have the disadvantage of converging to a set of probabilities which
is independent of prior assignments.

6. OUTLOOK

6.1 Open Problems

In this subsection some important problems, which are not yet
solved or not yet satisfactorily solved, are discussed. This is not
done by giving unsolved equations or not yet proved conjectures but
by stating unanswered questions.

A general problem in preprocessing is to find useful operations.
Although there are some intuitive and quantitative considerations
there is no general qualitative method to evaluate the power of preprocessing methods without designing and testing the whole image analysis system. If a complete system is available it is, of course,
possible to rank preprocessing methods according to success of analysis. The question is whether there are general criteria to measure
preprocessing quality without designing the whole system.

A similar problem exists for simple constituents of an image.
Again the question is whether there are general criteria to evaluate
the quality of simple constituents and whether there are general algorithms which, when applied to images of a particular field of

problems Ω, will generate a set of constituents (picture primitives) which is optimal in the sense of giving best success of analysis.

Representation and acquisition of structural information (or knowledge) is another important problem. It is not yet clear which the best method of representation is and which way of obtaining such information is the most efficient. It is also known that different levels of information - syntactic, semantic, and pragmatic - are extremely useful. A further question is to what extent these levels should be used and how to combine them most efficiently.

Some remarks about control structure were made. Again there is the question whether there are general algorithms to systematically generate that control structure which is suited most for a particular field of problems. It is questionable whether one structure (be it top-down or bottom-up, hierarchical or heterarchical, or something else) will be the best for all applications of image analysis.

The problem of developing an image-analysis system was divided into fairly independent subproblems, like preprocessing, structural information representation and so on. The question remains whether independent optimization of system components will finally lead to a globally optimal system. Another question is how to define optimality. Success of analysis, for instance, measured by error rates, certainly will be an important aspect; speed of analysis, storage requirements, and overall system costs are important, too. There does not seem to be an approach to these problems of system design.

The above problems follow from the previous sections and Fig.3. Some additional problems are depicted in Fig.5. A powerful, flexible, and "intelligent (?)" system should have capabilities beyond that of Fig.3. Generally, it should have the ability to improve its performance which is usually denoted as learning, and to maintain a certain degree of performance even if properties of input data change. There are applications where it may be useful if the system is not limited to knowledge about just one field of problems $\Omega^{(1)}$, but incorporates knowledge from other fields of problems $\Omega^{(2)},\ldots,\Omega^{(K)}$. An example would be a diagnosis system which uses thorax radiographs ($\Omega^{(1)}$), images obtained with radiopharmaceuticals ($\Omega^{(2)}$), and electrocardiograms ($\Omega^{(3)}$) to diagnose heart diseases.

These extensions pose the problems of automatically doing things which so far the designer of the systems does. It would require optimization of control module, development and adjustment of preprocessing and extraction methods, and inference of information (possibly from different fields of problems) by the system itself. It was mentioned that some methods of automatic and interactive inference of knowledge are already available. Certainly it will not be necessary for all kinds of applications to develop systems with this degree of sophistication.

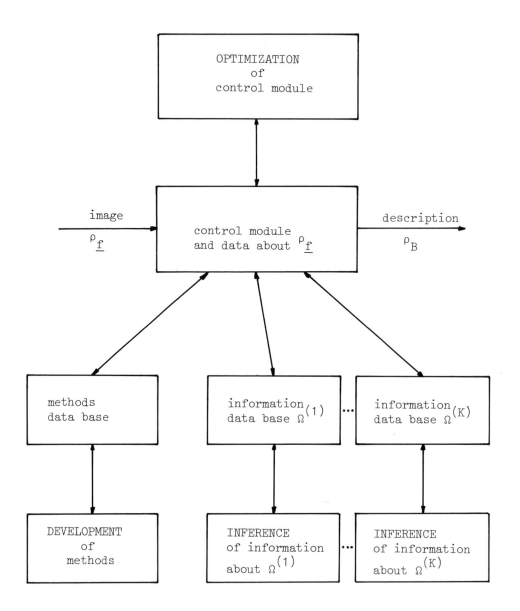

Fig. 5. An expanded system for imgage analysis.

Since so many problems in image analysis are still open, it seems appropriate to mention the following example. So far there is no generally optimal algorithm to play chess although it may exist. This does not prevent some people to play excellent chess. Equally there are some people who managed to design image analysis-systems which work satisfactorily in certain applications.

6.2 Applications

In the previous subsection we concentrated on some general research problems. It should not be overlooked that problems in applications may be different. Presently there are two applications of pictorial pattern classification and analysis, where commercial equipment is available: character recognition and blood-smear analysis. As pointed out in [6.1] the latter took a research effort of three decades and nevertheless is just at the point where these efforts are beginning to bear fruit. Past experience shows that quick progress is unlikely, but also that no principally unsolvable problems occurred. Therefore it is to be expected that image analysis (as well as classification and analysis of other patterns [6.2]) will become a routine tool in the fields mentioned in Section 1 and probably in fields not envisioned so far. To become a routine tool it will not only be necessary to develop more sophisticated methods of analysis but also get a more thorough understanding of existing ("standard") methods. This will be inevitable in order to develop systems which show high performance and are cost effective. In addition it is to be expected that achievements will be made which are inherent to the application (not to image analysis) but which greatly influence analysis techniques. A good example is the development of a chromosome staining method showing banding patterns [6.1]. With this, simple constituents can be extracted from the image which were "invisible" so far.

Presently, industrial and medical applications seem to be most promising since there is a great amount of pictorial information requiring routine examination. Increasing costs of labour make these fields interesting from an economical point of view. On the other hand, introduction of these techniques may give rise to social problems if people who presently do this work become unemployed.

6.3 Related Topics

Problems occurring in image analysis, like representation of information or control structure design, are common in other fields, too; see for instance [5.2,6.2-6.4]. The Hearsay II speech-understanding system is an example of a most interesting control structure allowing parallel operation of different modules. Since this paper is strictly limited to image analysis the field of speech recognition and understanding is only mentioned here. In this context, application of Viterbi algorithm to decode noisy symbol strings is of interest [6.5]. It is a method to assign a label (meaning of a word) to a string of simple constituents (phonemes) in the case of errors during extraction.

A specialized field of image analysis was not mentioned in this paper: analysis of sequences of images. Examples are analysis of x-ray angiograms or TV-images of traffic flow. These techniques allow tracing the time course of objects or patterns or processes. Surveys are given in [6.6, 6.7].

Search methods of problem solving may be valuable both in finding good components of an image analysis system and in obtaining the meaning of symbol strings. A general treatment of these techniques is given in [6.8]. A useful representation of asynchronous parallel activity is provided by Petri nets [6.10], which are in some aspects similar to flowchart schemes mentioned in Section 5.

Hardware aspects of image analysis were not considered here because this topic is covered by other papers of this volume. At least it seems that hardware components will not be any problem since components will be available in increasing numbers [6.9]. The problem seems to be how to find new applications for available hardware components (image analysis offers a rich potential), how to organize components (this is a problem in image analysis), and how to motivate people to pay for all these things. The last point certainly is of extreme importance for the further development of image-analysis applications, but fortunately it is not a point for a paper which was concerned with methods of digital-image analysis.

REFERENCES

[1.1] T. Pavlidis: Structural pattern recognition.
Springer Verlag, Berlin, 1977

[1.2] A. Rosenfeld: Picture processing: 1976
Computer Graphics and Image Processing 6, 157-183, 1977

[1.3] R. O. Duda, P. E. Hart: Pattern classification and scene analysis.
J. Wiley, New York, 1972

[1.4] H. Niemann: Methoden der Mustererkennung.
Akademische Verlagsgesellschaft, Frankfurt, 1974

[1.5] A. Rosenfeld (ed.): Digital picture analysis.
Springer Verlag, Berlin, 1976

[1.6] A. Rosenfeld, A. C. Kak: Digital picture processing.
Academic Press, New York, 1976

[1.7] K. S. Fu: Syntactic pattern recognition, applications.
Springer Verlag, Berlin, 1977

[1.8] A. Klinger, K. S. Fu, T. L. Kunii (eds.): Data structures, computer graphics, and pattern recognition.
Academic Press, New York, 1977

[1.9] A. R. Hanson, E. M. Riseman: Computer Vision Systems.
Academic Press, 1978

[1.10] H. Wechsler: Automatic detection of rib contours in chest radiographs. Birkhäuser Verlag, Basel, 1977, Band ISR 29

[1.11] C. H. Chen (ed.): Pattern recognition and artificial intelligence, Academic Press, New York, 1976, p.450

[1.12] K. S. Fu: Syntactic methods in pattern recognition.
Academic Press, New York, 1974

[2.1] H. G. Musmann: Digital image coding. See this volume.

[2.2] T. S. Huang, O. J. Tretiak: Picture bandwidth compression.
Gordon and Breach Science Pub., New York, 1972

[2.3] W. Schüssler: Digital filtering. This volume.

[2.4] T. S. Huang (ed.): Picture processing and digital filtering.
Springer Verlag, Berlin, 1975

[2.5] D. Casasent (ed.): Optical data processing.
Springer Verlag, Berlin, 1977

[2.6] J. W. Goodman: Introduction to Fourier-optics.
McGraw - Hill Book Comp., New York, 1968

[2.7] R. P. Kruger: Computer processing of radiographic images.
PhD dissertation, Eng. Dep., Univ. of Missouri, Columbia, 1971

[2.8] W. Frei: Image enhancement by histogram hyperbolization.
Computer Graphics and Image Proc. 6, 286-294, 1977

[2.9] W. Pratt: Digital image processing.
J. Wiley, New York, 1978

[2.10] R. B. Crane: Preprocessing techniques to reduce atmospheric and sensor variability in multispectral scanner data.
Proc. 7th Intern. Symp. on Remote Sensing of Environment, Univ. of Mich., Ann Arbor, Mich., 1345-1350, 1971

[2.11] K. H. Höhne, et al.: Functional images - a new tool for X-ray functional diagnostics.
DESY DV-78/01, Deutsches Elektronen-Synchrotron, Hamburg, 1978

[2.12]   D.F.Webber: Techniques for image registration.
         Proc.Conf.on Machine Processing of Remotely Sensed Data,
         Purdue Univ., 1973, (IEEE Cat.No.73 CHO 834-2GE)
[2.13]   S.S.Riffman: Digital rectification of ERTS multispectral
         imagery.
         Symp. on Sign. Results Obtained from ERTS, NASA SP-327,
         Goddard Space Flight Center, 1131-1142, 1973
[2.14]   A.R.Hanson, E.M.Riseman: The design of a semantically
         directed vision processor.
         COINS TR 75 c-1, Univ.of Mass., Amherst, 1975
[2.15]   I.S.Reed, et.al.: Image processing by transforms over a
         finite field.
         IEEE Trans.on Computers C-26, 874-881, 1977
[3.1]   L.S.Davis: A survey of edge detection techniques.
         Computer Graphics and Image Proc.4, 248-270, 1975
[3.2]   L.G.Roberts: Machine perception of three-dimensional solids.
         In: J.T.Tippelt et.al.(eds.): Optical and electro-optical
         information processing. The MIT-Press, Cambridge, Mass.,
         1965, 159-197
[3.3]   K.K.Pingle: Visual perception by a computer.
         In: A.Grasselli (ed.): Automatic interpretation and classi-
         fication of images, Academic Press, New York, 1969, 277-284
[3.4]   A.Rosenfeld: A nonlinear edge detection technique.
         Proc.IEEE 58, 814-816, 1970
[3.5]   M.Hueckel: An operator which locates edges in digitized
         pictures.
         JACM 18, 113-125, 1971
[3.6]   J.W.Modestino, R.W.Fries: Edge detection in noisy images
         using recursive digital filtering.
         Computer Graphics and Image Processing 6, 409-433, 1977
[3.7]   F.Holdermann, H.Kazmierczak: Generation of line drawings
         from grey-scale pictures.
         AGARD Conf.Proc.No.94 on Artificial Intelligence, 1970,
         814-816
[3.8]   R.O.Duda, P.E.Hart: Use of the Hough transformation to
         detect lines and curves in pictures.
         CACM 15, 11-15, 1972
[3.9]   C.Kimme, et.al.: Finding circles by an array of accumulators.
         CACM 18, 120-122, 1975
[3.10]   C.R.Brice, C.L.Fennema: Scene analysis using regions.
         Artificial Intelligence 1, 205-226, 1970
[3.11]   J.M.Prager, et.al.: Extracting and labelling boundary seg-
         ments in natural scenes.
         COINS Technical Report 77-7, Univ.of Massachusetts at Am-
         herst, 1977
[3.12]   G.J.Vander Brug: Experiments in iterative enhancement of
         linear features.
         Computer Graphics and Image Processing 6, 25-42, 1977

[3.13] S.W.Zucker, et.al.: An application of relaxation labeling to line and curve enhancement.
IEEE Trans.Comp.C-26, 394-403, 1977

[3.14] G.J.Vander Brug: Semilinear line detectors.
Computer Graphics and Image Processing 4, 287-293, 1975

[3.15] M.D.Kelly: Edge detection in pictures by computer using planning.
In: B.Meltzer, D.Michie (eds.), Machine Intelligence 6, Edinburgh, Univ.Press 1971, 397-409

[3.16] Y.Shirai: A context sensitive line finder for recognition of polyhedra.
Artificial Intelligence 4, 95-119, 1973

[3.17] R.Nevatia: A color edge detector.
Proc.3.Int.Joint Conf.on Pattern Recognition, Coronado, Calif., 1976, 829-832

[3.18] H.Freeman: On the encoding of arbitrary geometric configurations.
IRE Trans.El.Comp.EC-10, 260-268, 1961

[3.19] H.Freeman, J.M.Glass: On the quantization of line-drawing data.
IEEE Trans.Syst.Science and Cyb., SSC-5, 70-79, 1969

[3.20] C.T.Zahn, R.Z.Roskies: Fourier descriptors for plane closed curves.
IEEE Trans.Comp.C-21, 269-281, 1972

[3.21] G.H.Granlund: Fourier preprocessing for hand print character recognition.
IEEE Trans.Comp.C-21, 195-201, 1972

[3.22] E.Persoon, K.S.Fu: Shape discrimination using Fourier descriptors.
IEEE Trans.Syst., Man, and Cyb. SMC-7, 170-179, 1977

[3.23] M.K.Hu: Visual pattern recognition by moment invariants.
IRE Trans.Inf.Th., IT-8, 179-187, 1962

[3.24] S.Dudani, et.al.: Aircraft identification by moment invariants.
IEEE Trans.Comp. C-26, 39-46, 1977

[3.25] S.W.Zucker: Region growing: childhood and adolescence.
Comp.Graphics and Image Processing 5, 382-399, 1976

[3.26] E.M.Riseman, M.A.Arbib: Computational techniques in the visual segmentation of static scenes.
Computer Graphics and Image Processing 6, 221-276, 1977

[3.27] R.S.Ledley: High-speed automatic analysis of biomedical pictures.
Science 146, 216-223, 1964

[3.28] J.Schürmann: Bildvorbereitung für die automatische Zeichenerkennung.
Wiss.Berichte AEG-Telefunken, 1974, 90-99

[3.29] J.L.Muerle, D.C.Allen: Experimental evaluation of techniques for automatic segmentation of objects in a complex scene.
In: G.C.Cheng, et.al.(eds.), Pictorial Pattern Recognition, Tompson, Washington, 1968, 3-13

[3.30]   T.V.Robertson, et.al.: Multispectral image partitioning.
         TR-EE 73-26, School of El.Eng., Purdue Univ., 1973
         (LARS Inf.Note o71373)

[3.31]   T.Pavlidis: Segmentation of pictures and maps through
         functional approximation.
         Comp.Graphics and Image Processing 1, 360-372, 1972

[3.32]   S.L.Horowitz, T.Pavlidis: Picture segmentation by a directed
         split - and - merge procedure.
         Proc.2.Int.Joint Conf.on Pattern Recognition, Copenhagen,
         1974, 424-433

[3.33]   Y.Yakimovsky, J.Feldman: A semantics - based decision
         theory region analyzer.
         Proc.3.Int.Joint Conf.on Art.Int. 1973, 580-588

[3.34]   J.M.Tenenbaum, H.G.Barrow: Experiments in interpretation
         guided segmentation.
         Artificial Intelligence 8, 241-274, 1977

[3.35]   P.Brodatz: Textures. Dover, New York, 1966

[3.36]   B.Julesz: Experiments in the visual perception of texture.
         Scientific American vol.232, No.4, 34-43, 1975

[3.37]   J.S.Weszka, et.al.: A comparative study of texture measures
         for terrain classification.
         IEEE Trans.on Syst., Man, and Cyb., SMC-6, 269-285, 1976

[3.38]   J.K.Hawkins: Textural properties for pattern recognition
         In: B.S.Lipkin, A.Rosenfeld (eds.): Picture Processing
         and Psychopictorics. Academic Press, New York, 1970, 347-370

[3.39]   R.M.Haralick, et.al.: Textural features for image classifi-
         cation.
         IEEE Trans.on Syst., Man, and Cyb., SMC-3, 610-621, 1973

[3.40]   E.S.Deutsch, N.J.Belknap: Texture descriptors using neigh-
         borhood information.
         Computer Graphics and Image Processing 1, 145-168, 1972

[3.41]   J.T.Tou, Y.S.Chang: Picture understanding by machine via
         textural feature extraction.
         Proc.IEEE Conf.on Pattern Recognition and Image Processing,
         Troy, 1977, 392-399

[3.42]   M.M.Galloway: Texture analysis using grey level run lengths.
         Computer Graphics and Image Processing 4, 172-179, 1975

[3.43]   S.Tsuji, F.Tomita: A structural analyzer for a class of
         textures.
         Computer Graphics and Image Processing 2, 216-231, 1973

[3.44]   S.W.Zucker, et.al.: Picture segmentation by texture dis-
         crimination.
         IEEE Trans.on Comp.C-24, 1228-1233, 1975

[3.45]   J.P.Foith: Symbolische Repräsentationen von Grauwertbildern
         für Szenenanalysen.
         Dissertation, Univ.Erlangen-Nürnberg, 1978

[3.46]   O.R.Mitchell, et.al.: A max-min measure for image texture
         analysis
         IEEE Trans.on Comp.C-26, 4o8-414, 1977

[3.47]   J.T.Tippelt, et.al.(eds.): Optical and electrooptical infor-
         mation processing.
         The MIT Press, Cambridge, Mass., 1965, 199-2o7 and 535-550
[3.48]   D.G.Olsen: The growing range of multichannel detection.
         Optical Spectra, Feb.1971
[4.1]   Y.C.Ho, A.K.Agrawala: On pattern classification algorithms -
         introduction and survey.
         IEEE Trans.on Aut.Control AC-13, 676-690, 1968
[4.2]   K.Fukunaga: Introduction to statistical pattern recognition.
         Academic Press, New York, 1972
[4.3]   J.Schürmann: Polynomklassifitatoren für die Zeichenerkennung.
         R.Oldenbourg, München, 1977
[4.4]   G.Winkler: Stochastische Systeme - Analyse und Synthese.
         Akademische Verlagsgesellschaft, Wiesbaden, 1977
[4.5]   P.Haberäcker: Multispektrale Klassifizierung in DIBIAS.
         In: H.H.Nagel (ed.): Digitale Bildverarbeitung.
         Informatik Fachberichte 8, Springer Verlag, Berlin, 1977, 71-79
[4.6]   G.J.Vander Brug, A.Rosenfeld: Two - stage template matching.
         IEEE Trans.on Computers C-26, 384-393, 1977
[4.7]   M.A.Fischler, R.A.Elschlager: The representation and matching
         of pictorial structures.
         IEEE Trans.on Computer C-22, 67-92, 1973
[4.8]   R.Y.Wong, E.L.Hall: Sequential hierarchical scene matching.
         IEEE Trans.on Computers C-27, 359-366, 1978
[4.9]   A.K.Dewdney: Analysis of a steepest-descent image-matching
         algorithm.
         Pattern Recognition 10, 31-39, 1978
[4.10]   T.M.Cover, P.E.Hart: Nearest neighbour pattern classification.
         IEEE Trans.on Inf.Theory IT-13, 21-27, 1967
[4.11]   A.V.Kulkarni, L.N.Kanal: An optimization approach to hierarchi-
         cal classifier design.
         Proc.3.Int.Joint Conf.on Pattern Recognition, Coronado, 1976,
         459-466
[4.12]   R.M.Ray, et.al.: Implementation of ILLIAC 4 algorithms for
         multispectral image interpretation.
         Center of Advanced Computation, Univ.of Urbana, Ill., 1974
[4.13]   G.H.Ball, D.J.Hall: A clustering technique for summarizing
         multivariate data. Behavioral Science 12, 153-155, 1967
[5.1]   M.A.Arbib, E.M.Riseman: Computational techniques in visual
         systems.
         COINS Technical Report 76-10, Univ.of Mass. at Amherst, 1976
[5.2]   R.Davis, et.al.: Production rules as a representation for a
         knowledge based consultation program.
         Artificial Intelligence 8, 15-45, 1977
[5.3]   L.D.Erman, V.R.Lesser: System engineering techniques for
         artificial intelligence systems.
         Dep.of Comp.Science, Carnegie-Mellon Univ., 1977

[5.4] O. Akin, R. Reddy: Knowledge acquisition for image under-
        standing research.
        Computer Graphics and Image Processing 6, 307-334, 1977
[5.5] R. D. Fennell: Multiprocess software architecture for AI
        problem solving.
        PhD dissertation, Dep. of Comp. Science, Carnegie-Mellon
        Univ., Pittsburgh, 1975
[5.6] A. Newell: Production systems; models of control structures.
        In: W. C. Chase (ed.): Visual Information Processing, Academic
        Press, New York, 1973, 463-526
[5.7] M. Minsky: A framework for representing knowledge.
        AI Memo 306, Mass. Inst. of Technology, 1974
[5.8] Z. Manna: Properites of programs and the first-order predi-
        cate calculus.
        JACM 16, 244-255, 1969
[5.9] Z. Manna: Mathematical theory of computation.
        McGraw-Hill Book, New York, 1974
[5.10] C. A. Harlow: Image analysis and graphs.
        Computer Graphics and Image Processing 2, 60-82, 1973
[5.11] J. E. Hopcroft, J. D. Ullmann: Formal languages and their re-
        lation to automata.
        Addison-Wesley Publ. Comp., 1969
[5.12] M. Gross, A. Lentin: Mathematische Linguistik.
        Springer Verlag, Berlin, 1971
[5.13] A. V. Aho, J. D. Ullmann: The theory of parsing, translation,
        and compiling.
        Prentice Hall, Englewood Cliffs N. J., 1972
[5.14] H. J. Schneider: Compiler, Aufbau und Wirkungsweise.
        De Gruyter Verlag, Berlin, 1975
[5.15] D. J. Rosenkrantz: Programmed grammars and classes of formal
        languages.
        JACM 16, 107-131, 1969
[5.16] S. Y. Lu, K. S. Fu: Stochastic error-correcting syntax analysis
        for recognition of noisy patterns.
        IEEE Trans. on Computers C-26, 1268-1276, 1977
[5.17] K. S. Fu, B. K. Bhargava: Tree systems for syntactic pattern
        recognition.
        IEEE Trans. on Computers C-22, 1087-1099, 1973
[5.18] S. M. Chou, K. S. Fu: Transition networks for pattern recognition.
        Technical Report TR-EE 75-39, Purdue Univ., 1975
[5.19] D. C. Milgram, A. Rosenfeld: Array automata and array grammars.
        Proc. IFIP Congress 1971, 166-173
[5.20] A. Rosenfeld, D. C. Milgram: Web automata and weg grammars.
        In: B. Meltzer, D. Michie (eds.): Machine Intelligence 7,
        J. Wiley, 1972, 307-324
[5.21] M. Nagl: Formale Sprachen von markierten Graphen.
        Arbeitsbericht des IMMD 7, Nr. 4, Univ. Erlangen-Nürnberg, 1974
[5.22] K. S. Fu, T. L. Booth: Grammatical inference - introduction and
        survey. IEEE Trans. on Systems, Man, and Cyb. SMC-5, 1975,
        95-111 and 409-423

[5.23] M. Yachida, S. Tsuji: A versatile machine vision system for
        complex industrial parts.
        IEEE Trans. on Computers C-26, 882-894, 1977

[5.24] R. S. Ledley, et. al.: FIDAC film input to digital automatic
        computer and associated syntax directed pattern recognition
        programming.
        In [3.47]

[5.25] M. G. Thomason, R. C. Gonzalez: Syntactic recognition of imper-
        fectly specified patterns.
        IEEE Trans. on Computers C-24, 93-95, 1975

[5.26] A. C. Shaw: A formal picture description scheme as a basis
        for picture processing systems.
        Information and Control 14, 9-52, 1969

[5.27] A. C. Shaw: Parsing of graph - representable pictures.
        JACM 17, 453-481, 1970

[5.28] E. H. Sussenguth: Structure matching in information processing.
        PhD Thesis, Appl. Math., Harvard Univ., Cambridge, Mass., 1964

[5.29] M. Sties: Analysis and symbolic description of pictures and
        localization of objects.
        Proc. Conf. on Machine Perception of Patterns and Pictures,
        Teddington, 1972, Paper 29

[5.30] E. C. Frender: Structural isomorphism of picture graphs.
        In [1.11]

[5.31] H. G. Barrow, et. al.: Some techniques for recognising structures
        in pictures.
        In: S. Watanabe (ed.): Frontiers of Pattern Recognition, Aca-
        demic Press, New York, 1972, 1-29

[5.32] G. Tinhofer: Methoden der angewandten Graphentheorie.
        Springer Verlag, Wien, 1976

[5.33] A. Rosenfeld, et. al.: Scene labeling by relaxation operations.
        IEEE Trans. on Systems, Man, and Cyb. SMC-6, 420-433, 1976

[5.34] A. Rosenfeld: Iterative methods in image analysis.
        Proc. IEEE Conf. on Pattern Rec. and Image Processing, Troy,
        1977, 14-18

[5.35] J. M. Tenenbaum, H. G. Barrow: IGS: A paradigm for integrating
        image segmentation and interpretation.
        In [1.11], 472-507

[5.36] S. Peleg, A. Rosenfeld: Determining compatibility coefficients
        for relaxation processes.
        Techn. Report TR-570, Comp. Science Center, Univ. of Maryland,
        1977

[5.37] G. C. Nicolae, K. H. Höhne: Digital video system for real-time
        processing of image series.
        DESY DV-78/02, Deutsches Elektronen-Synchrotron, Hamburg, 1978

[6.1] K. Preston: Digital picture analysis in cytology.
        In [1.5], 209-294

[6.2] D. R. Reddy: Speech recognition by machine, a review.
        Proc. IEEE 64, 501-531, 1976

[6.3]   V. R. Lesser, et. al.: Organization of Hearsay II speech under-
         standing system.
         IEEE Trans. on Acoustics, Speech, and Signal
         Proc. ASSP-23, 11-24, 1975

[6.4]   R. Fennell, V. R. Lesser: Parallelism in artificial intelli-
         gence problem solving; a case study of Hearsay II.
         IEEE Trans. on Computers C-26, 98-111, 1977

[6.5]   L. R. Bahl, et. al.: Decoding for channels with insertions,
         deletions, and substitutions with applications to speech
         recognition.
         IEEE Trans. on Inf. Theory IT-21, 404-411, 1975

[6.6]   H. H. Nagel: Analysis techniques of image sequences.
         Proc. 4. Int. Joint Conf. on Pattern Recognition, Kyoto, Japan,
         1978

[6.7]   Martin, Aggarwal: Dynamic scene analysis.
         Computer Graphics and Image Processing 7, 356-374 (1978)

[6.8]   N. J. Nilsson: Problem-solving methods in artificial intelli-
         gence.
         McGraw-Hill Book, New York, 1971

[6.9]   N. Lindgren: Semiconductors face the '80s.
         IEEE Spectrum 14, No. 10, 42-48, Oct. 1977

[6.10]   J. L. Peterson: Petri nets.
         Computing Surveys 9, No. 5, 223-252, 1977

APPLICATION

BIOMEDICAL IMAGE PROCESSING

K. Preston, Jr.
Carnegie-Mellon University
Pittsburgh, Pennsylvania
U. S. A.

INTRODUCTION

This chapter discusses electronic image processing in medicine. Electronic image processing includes: (1) electronic display; (2) automatic display enhancement; (3) automatic image analysis. The word 'analysis' implies operations which include image partitioning, object detection, and object identification. Although this chapter is primarily concerned with item (3), items (1) and (2) are briefly discussed.

The history of electronic image processing in biomedicine begins, of course, with television. As is shown in Figure 1, commercial television in the United States began with the installation of the first television stations in 1940. Each television transmitter produces about 10^{14} picture elements or 'pixels' per year. In the United States, the number of pixels per year generated by commercial television stations climbed rapidly during the interval 1945 - 1950 and then stabilized at a level of about 7×10^{16} pixels per year by the 1960s [1]. This, of course, is broadcast output and does not include the 10^8 television receivers now in use in the United States.

Electronic imaging has permeated the entire world in the short span of 30 years. As shown in Figure 1, the world is now entering a new era of sudden change in two fields of great interest in biomedicine, namely, the automated analysis of images of human white-blood cells (white-cell differential) and in the generation and analysis of images using computerized tomography. It appears (Figure 1) that the number of pixels per year generated and processed in these areas will

125

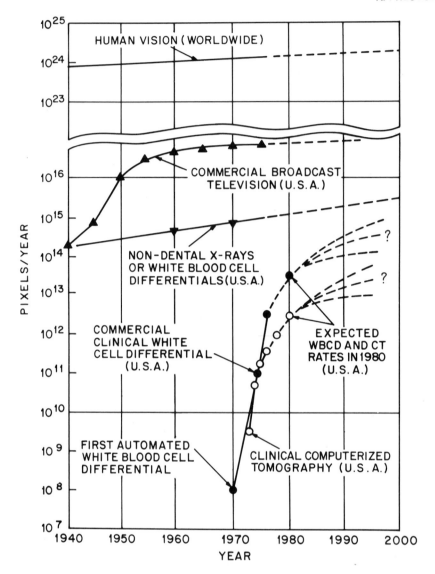

Figure 1. Graph showing the relationship between image sensing (human vision), electronic image generation (commercial television), and digital image processing (computerized tomography and the white blood-cell differential count) during the 20th century. A base line is provided showing the rate of visual image processing of x-rays and blood cells. The perception rate of human vision is based on 10^6 pixels per eye, a frequency cutoff at 10 Hz, and an average usage 16 hours/day using the von Foerster equation for world population:

$$N = 1.79 \times 10^{11}/(2026.87 - t)^{0.99}.$$

grow by five orders of magnitude during the decade of the 1970s. In addition, there appears to be gathering interest in using electronic image analysis for processing nuclear images and image constructing using ultrasound. These developments are discussed in more detail below.

Work in the field of biomedical image display and enhancement began with the work of Land [2] and Zworykin and Hatke [3] in the late 1940s and the early 1950s. Their contribution was the electronic color-translating microscope which transformed and enhanced ultraviolet images (at 265, 285, 315 nm) into the primary colors of a television monitor. With this system, it was possible to observe the streaming of cellular protoplasm, the cyclosis of chloroblasts and other such phenomena in real time.

Work in automatic electronic image analysis began at the same time in Great Britain when a committee of the National Coal Board was urged by Dr. Jocob Bronowski to investigate 'the possibility of making a machine to replace the human observer'. This is described in Walton [4]. The driving force behind this effort was the desire to quantitate the size and number of coal-dust particles on microscope slides and, by the mid-1950s, the so-called 'flying-spot' microscope of Causley and Young [5] was in use counting and sizing red cells at 1 μm resolution and generating a histogram of particle sizes (0 - 30 μm in 5 μm intervals) in a 500 × 500 μm field in four seconds. Agreement between machine and visual observation of the same field was reported to be within 2 %. Commercialization of this apparatus took place in the 1950s by various concerns, such as Mullard and C.F. Casella, Ltd.

It was not until the 1960s that the first successful application of automatic electronic image analysis occurred in biomedicine. The United States Atomic Energy Commission began a program in 1960 at the Perkin-Elmer Corp. which led to the fabrication of the CELLSCAN system [6, 7] for the purpose of doing research in the automatic analysis of white blood-cell images. This project was later funded in turn by the National Institute of Health and by the Department of Defense finally leading to the successful solution of the problem as reported in [8]. The project represented the first (and only) successful embodiment in hardware of a real-time biomedical image analysis system. Other contemporary efforts were begun in the analysis of chest x-rays [9], chromosomes [10], and Papanicolau smears [11]. These efforts have not led to commercial success.

The 1970s saw the exciting development of x-ray tomography by Hounsfield [12] in Great Britain which led to the rapid commercialization of x-ray tomography for brain scanning and, later, for whole-body scanning. It is estimated that there will be over 1000 x-ray computerized tomography machines in the United States by 1980 having an output of approximately 5×10^{12} pixels per year. At the same time, the CELLSCAN project at Perkin-Elmer was commercialized in the form of the 'diff3 analyzer' which became a commercial product in 1977 and is now to be manufactured by Coulter Electronics Co. Other United States corporations have entered this field, namely, Corning Glass Works [13], SmithKline Corp. [14], and Abbott Laboratories [15]. The specifications of these image-processing systems for analyzing the images of human white blood cells are given in Table 1. Figure 1 indicates a growth of the order of 5×10^{13} pixels per year for these machines (approximately 100 have been sold in the United States to date) by the year 1980.

NUCLEAR-MEDICINE IMAGING

Image processing in nuclear medicine is still in its infancy but this chapter would be incomplete without mentioning of methodology. The patient is rendered radioluminescent by the injection of tracer amounts of one or more radiopharmaceuticals. The image of the patient is then formed by means of a scanning gamma-camera. This device consists of a focused collimator made of a heavy metal penetrated by holes which produce focusing over a small volume within the patient. Gamma radiation from within the patient passes through these holes to a large area detector. The most frequently used detector material is sodium iodide doped with small amounts of thallium. This material converts the gamma radiation into visible light photons in an array of phototubes which are in turn connected to an appropriate electronic network whose purpose (in conjuntion with the collimator) is to locate the source of the gamma radiation within the patient. This entire assembly is raster-scanned over the patient and a recorder produces an image such as that shown in Figure 2. A general description of nuclear-medicine imaging technology may be found in [16] as well as in articles published in the IEEE Transactions on Nuclear Science.

Table 1. Characteristics of the four commercial clinical white blood-cell differentiating instruments now manufactured in the United States.

	ADC 500	dIff3	Hematrak®	LARC®
SYSTEM				
SOURCE	J. E. Green	G. T. Paul	M. N. Miller	C. H. Rogers
AFFILIATION	Abbott Laboratories	Perkin-Elmer	Geometric Data	Corning
OPTICS				
OBJECTIVE				
Magnification	40	40	40	100
Numerical Aperture	1.0	1.0	1.0	1.3
Type	Planachromat	Planachromat	Planachromat	Planachromat
Immersion Medium (Index)	Oil (1.518)	Oil	Oil (1.518)	Oil (1.515)
OCULAR				
Magnification	10	12.5	10	10
Type	Widefield	KPL	KPL	(Adjustable Focus)
ILLUMINATION				
Light Source	150w Zenon	60w Tungsten	CRT	80w Tungsten
Wavelength(s) (nm)	412, 525, 560	510, 580	420, 520, 650	Orange, Blue-Green
Bandwidth(s) (nm)	30, 20, 20	30, 20	-	-
MODULATION TRANSFER				
FUNCTION	-	-	-	-
SCANNERS				
MECHANICAL				
Slide Magazine Capacity	50	14	1	1
Stage Increments (X,Y,Z) (μm)	8, 8, -	3, 3, 0.1	350, 350, 0.25	5, 5, 0.1
Increment Time (X,Y,Z) (ms)	0.1-0.5, 0.1-0.5, -	1, 2, 2	70(X,Y)	2, 2, 3
Total Travel (X,Y,Z) (mm)	≈80, ≈80, 0.1	20, 10, 0.5	-	60, 20, 0.25
SEARCH MODE				
Type of Scanner	50 x 50 Diode Array	Linear Diode Array	CRT	Mirror
Scan Lines/Sec.	3,000	1,000	20,000	500
Spot Velocity (μ/ms)	2,500	-	-	150
Line Length (μm)	400	200	300	≈300
Line Width (μm)	8	3	300	≈7
Line Spacing (μm)	8	3	-	≈5
Coverage (mm²/min.)	600	36	-	45
Focus Method	Dual 128-Diode Scanner	Not Used	Max. High Video Freq.	Move Objective
Focus Response (sec.)	0.01	N/A	-	Real Time
Cell Acquisition Time (sec.)	0.016	0.15	0.05	0.10
Video Bandwidth	-	75kHz	-	-
IMAGING MODE				
Type of Scanner	Three 50 x 50 Diode Arrays	Plumbicon	CRT	Plumbicon
Pixel Spacing (μm)	0.5	0.4	0.25	0.4
Acquisition Rate (pixels/sec.)	150,000	$2.5 \cdot 10^6$	-	-
Acquisition Time (sec.)	0.016	0.02	-	-
Format (x pixels, y pixels)	50 x 50	128, 128	112, 80	48, 48
Video Bandwidth	-	1 MHz	-	-
Bits/Pixel	6	6	-	6
IMAGE PROCESSOR				
Type of Computer	Nova-2 and Special	Mini and Special	HP2105	PDP 8M
Image Memory (words)	-	16 - 20 k	-	16k
Operating System (words)	12K	32 k	8k	-
Measurements Made	8(Red Cell)/8(White Cell)	20(Red Cell)/50(White Cell)	96	8
Measurement Time (ms)	16	500	Real Time	-
Recognition Time (ms)	50	7	30	-
Processing Rate (cells/hr.)	29,000 (Av.) 36,000 (Peak)	3,500 - 4,000 (Av. White) 5,000 - 6,000 (Av. Red)	12,000 (peak)	5,500 (Av.) 8,000 (peak)
Output	Printed Lab. Ticket; Computer Link	Printed Lab. Ticket; Computer Link	Printed Lab. Ticket	Printed Lab. Ticket; Computer Interface

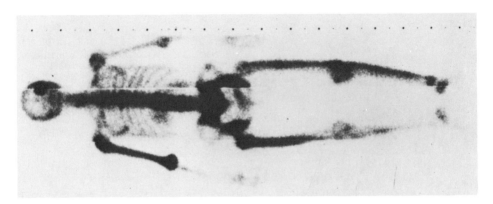

Figure 2. Bone image of a 60 year-old female with extensive Paget's disease, obtained by using Tc-99m MDP and a scintillation camera with moving table. Left skull, right humerus, spine, pelvis, both femurae and left tibia show increase uptake (courtesy Dr. Robert S. Hattner, University of California, San Francisco).

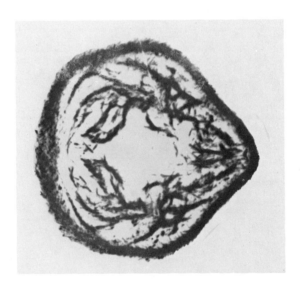

Figure 3. An ultrasonic quasi-tomogram of the human neck constructed using electronic analog techniques. This image was formed from projections taken in a pulse-reflection mode using a single ultra-sonic source-detector scanned over a full 360 degree arc (in water) around the neck of the patient. Thousands of reflections were summed via a cathode ray tube on photographic film over a period of several minutes. (Courtesy the late Dr. D. Howry, Denver, Colorado, U.S.A.)

ULTRASOUND IMAGING

The pulse-echo ultrasound scanner forms an image from echoes received from the interior structures of the human body. If desired, each echo may be displayed on a oscilloscope as a waveform. This is in contradistinction to x-ray and nuclear-medicine scanners which provide no waveform at all but instead integrate the transmission of radiation passing rectilinearly through various parts of the body. As early as the 1950s [17], this provided researchers with a 'third dimension' which x-ray could not provide until the invention of computerized tomography (see below).

Figure 3 shows an ultrasonogram which is a cross-section of the human neck produced by immersing the patient in water and rotating an ultrasonic transducer-detector around the patient in a circle. The received echoes were summed on either film or on a storage-cathode ray tube to produce the final image. More recent ultrasound scanners produce a multiplicity of images (Figure 4) again using a water-bath rectilinear scanner which is indexed by a few millimeters after each scan so that an entire volume in the patient is recorded. Figure 4 illustrates an advanced system which generates images at the rate of 50 per minute. Since each image consists of approximately one million data points, this system, considered as a digital-data source, produces about one megabyte per second or about 50 megabytes per minute. Because of the high data-generation rates involved, electronic image analysis (other than enhancement techniques for display) are yet to be applied to ultrasound imaging.

TOMOGRAPHIC IMAGING

Several workers [18 - 20] produced a breakthrough in x-radiology in the early 1970s. Their invention was the computerization of tomography and led to the introduction of the computerized tomography (CT) scanner produced by EMI Ltd. in 1972. This scanner used a single x-ray source paired with a detector which executed 180 linear scans at one-degree increments in approximately five minutes. During this time, tens of thousands of readings from the detector were digitally recorded and, from these readings, the image was reconstructed.

Since the source and detector scan on opposite sides of the patient, each linear scan produces a projection of the radiographic transmission values of the body along a line parallel to the scan direction. Therefore, the image formed from a multiplicity of these projections is

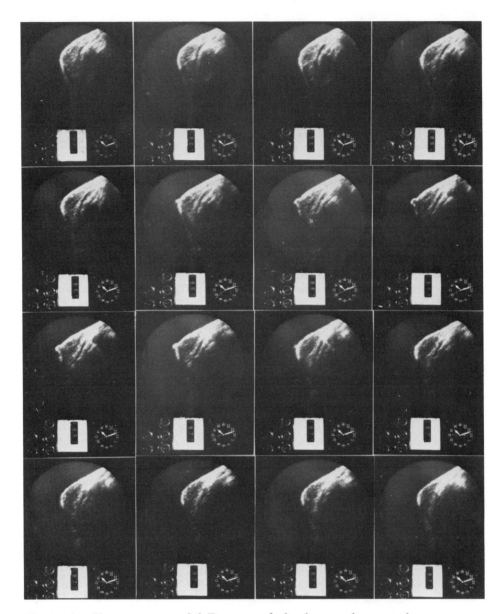

Figure 4. Sixteen sequential B-scans of the human breast taken at approximately one-second intervals at different lateral positions (courtesy Dr. G. Baum, Department of Ophthamology, Albert Einstein College of Medicine).

called a 'linear reconstruction from multiple projections'. The basic principle of forming the tomographic image is illustrated in Figure 5. If only three projections are combined, the image is relatively imperfect. If many projections are combined, an intelligible image is formed. The only problem is a background 'halo'. The value of this halo falls off radially at a rapid rate and is a known function. Thus, the tomographic image may be 'deconvolved' using the inversion of this known function as a correction factor. This operation is carried out by computer. Hence, the name 'computerized tomography'. Inaccuracies in the scan and computational problems in correcting the image have led to extensive studies of image reconstruction algorithms [21].

Progress in CT imaging has been rapid since 1972. Besides work on improving reconstruction algorithms, a major effort was applied to shortening the time required for scanning. During the relatively long scanning time of the original CT instruments, motion of the patient contributed to significant image blur. It was found that the most economic attack on the problem was to use a fan beam of x-radiation which was simultaneously detected by an array of detectors. CT scanners based on this principle are now available wherein the scan is produced in a few seconds. At the same time, the associated computer has been re-designed and specialized to generate the reconstructed image in real time.

A scan time of a few seconds is satisfactory in arresting all motions except that of the heart. Therefore, research continues towards even higher speed machines. One of the more ambitious of these programs is that now being carried out at the Mayo Foundation in Rochester, Minnesota [22]. This effort will lead to the construction of a new system to be called the Dynamic Spatial Reconstructor (DSR) which is intended to provide dynamic, stop-action, three-dimensional reconstructions of any part of the body. The DSR will sense and digitize at 60 frames per second 28 simultaneous planar projections. These projections will be created by 28 video-scanned, pulse x-ray source-detectors arranged in a semi-circle. Four-to-one multiplexing will be used to combine the outputs of seven sets of four video cameras which will feed the four video disks to be used for intermediate storage. A CDC 3500 computer complex will be used for image reconstruction. The special high-speed computer interface which will be required is described in [23] which mentions that video digitization will be by 9-bit analog/digital convertors operating in the range of 20 - 40 megasamples per second. This is planned to yield a total data-transfer rate of 500 megabits per second. Some recent results generated using a portion of this system are shown in Figure 6 which shows several images of the intact-working heart of a laboratory animal. Using advanced computer-graphics techniques (three-dimensional boundary

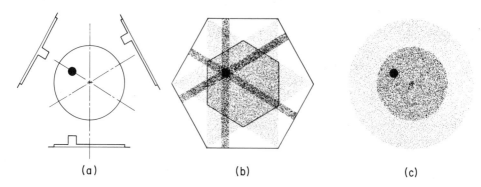

(a)   (b)   (c)

Figure 5.   The principle of tomographic reconstruction is to produce a planar image from its linear projections. Three projections (at 120 degree angles) are shown (a) with the image reconstructed from these projections (b). When using only three projections, the boundaries of the two regions in the image are hexagonal and there is much background noise. Through the use of the large number of projections. (c) image quality is improved.

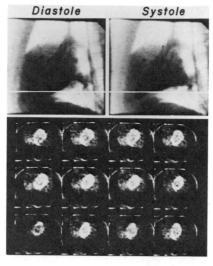

Figure 6.   X-ray video projection images of beating heart in intact dog (top) recorded at end-diastole and end-systole during infusion of x-ray contrast material into left ventricle and cross-section of intact-beating heart (bottom) reconstructed at the level of the brightened line for 12 points in time throughout the cardiac cycle. (Reproduced with permission from Robb *et al.* [22].)

surface detection, hidden surface removal, and shading) a computer-generated display of the whole isolated heart may be generated (Figure 7). Note that the bottom panel of this figure is a display of the same heart mathematically divided into halves to permit viewing of the structural detail of the interior.

BLOOD-CELL IMAGE ANALYSIS

Human white-blood cells (Figure 8) may be recognized using both colormetric and morphological features. The diff3 white blood-cell analyzer (mentioned above) first finds and focuses the image of an individual white blood cell in the blood smear on the microscope slide. Two images of the cell are then digitized using first yellow-red illumination and then green illumination. This data is stored in two gray-level image memories at six bits per pixel. The diff3 employs the Golay Transform [24, 25] in making all image measurements. For shape-dependent measurements (such as indentations in the nucleus, number of nuclear lobes, nuclear-to-cytoplasm area ratio, and cell perimeter) the analysis is performed directly using sequences of Golay Transforms. In the case of measurements relating to cell-constituent transmission (or optical density), the Golay Transform is used indirectly. Since diff3 uses the transform initially to identify cell-component parts (nucleus, cytoplasm, etc.), templates have been produced for each part and are resident in the diff3 image processor. These templates are used for the determination of the average optical density of the pixels which comprise these component parts. Finally, diff3 combines the morphological pattern features and the photometric pattern features into a pattern vector which is classified by hyperplane partitioning techniques to identify the cell type. In the overall process which is carried out, the Golay Transform is central to all pattern measurement and classification operations and provides the basis for the decisions of the diff3 blood-cell classifier.

Figure 9 is a 5 × 9 matrix of pictures representing various stages of white blood-cell image processing using the diff3. Each column relates to a different type of cell and each row represents a different stage of the processing for that particular cell type. The first row shows the gray-level representation of the scene generated by the high-resolution television microscope. The field of view represented is approximately 30 micrometers square. Each picture is spatially sampled on a 64 × 64 point grid with each point quantized to 64 levels so as to provide six bits of gray-level information. In each case, the white-blood cell to be analyzed is centered in the field with some red-blood cells and, sometimes, platelets also being present.

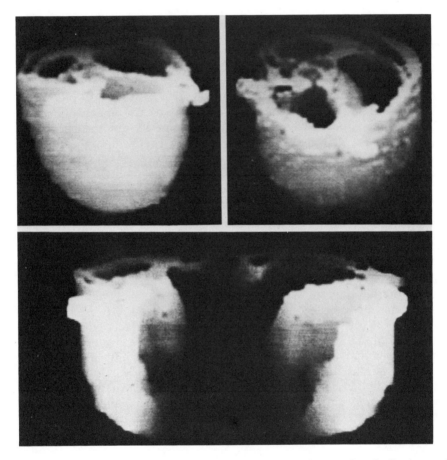

Figure 7. Computer generated three-dimensional gray-level displays of isolated intact dog heart determined from 30 reconstructed cross-sectional images extending from base-to-apex of heart. (Reproduced with permission from Robb *et al.* [22].)

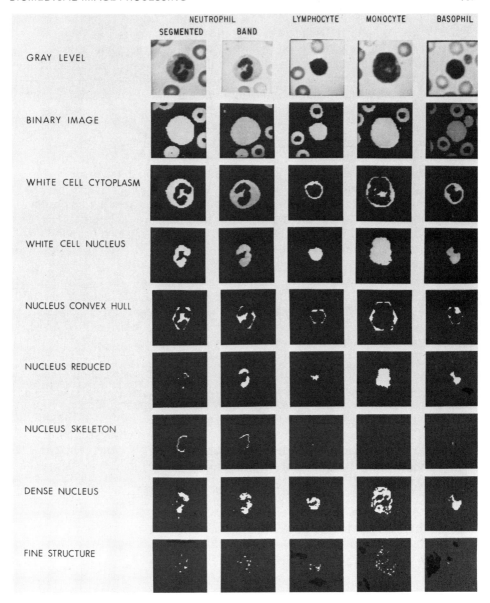

Figure 9. Detailed pictorial of the major intermediate results obtained by the diff3 Golay Transforms used for extracting features of the human white blood-cell image in the process of generating a white blood-cell differential count. Input images (green illumination) are shown in the top row. The lower rows show a multiplicity of one-bit 64 × 64 images produced at certain stages of the 500-step sequence carried out by diff3 in approximately 300 milliseconds. (Courtesy Mr. P. E. Norgren, Perkin-Elmer Corp., Norwalk, CT, U.S.A.)

The images shown in each row below the gray-level pictures in Figure 9 are all one-bit images having the same scale and a 64 × 64 format. They illustrate some of the morphological processing steps used to measure cell characteristics. Although the results of all measurement steps in the entire sequence are not shown, these pictures give a representative sample of some important measurements produced in the process of doing a white-cell differential count.

The second row of images shows the one-bit image extracted from the corresponding gray-level image by thresholding at a transmission level selected so as to separate the background from all cytological constituents in the field. Note that other objects are present in this field along with the white cell. Some of these touch the border of the field or the white cell while some of them are isolated. In the case of the basophil, for example, a large platelet near the basophil appears jointed to the basophil.

The third row shows the step at which the white-cell cytoplasm, a significant cellular constituent, has been extracted. The central object in the field (the desired white cell) has been isolated from all other objects. In the case of the basophil, the platelet has been removed. The nucleus and cytoplasm of the cell have been separated.

The nucleus itself, which is another significant constituent, is shown in the fourth row. Note that, in the case of the monocyte, the nucleus is shown as solid, although simple thresholding would leave internal voids in the nuclear image (see row 3). The sequence which produces the filled image of the nucleus uses a propagating function which starts at the image border and is arrested at the outer nuclear edge. (Propagating functions are described elsewhere in this book in the chapter by Duff).

The fifth row shows one-bit images in which the pseudo-convex hull of the nucleus has been generated. This operation (along with the generation of partial convex hulls) is useful in measuring concavities of various sizes in the nuclear silhouette.

The sixth row shows the result of reducing the nucleus using a sequence of shrinking operations. The area and perimeter measures of the reduced nucleus have been found, empirically, to be significant in cell indentification [26]. The examples of segmented and banded neutrophils shown in Figure 9 were taken deliberately to illustrate this point. In the upper row, they seem to be somewhat equivocal in their morphology. The gray-level image of the segmented neutrophil looks as if it has a continuous nucleus but its reduced nucleus shows a significant difference in comparison with the reduced nucleus of the band.

The seventh row of the figure shows the nuclear skeleton. The number of pixels in the skeleton is used as another morphological measure. The skeleton is then reduced to residues for the purpose of counting the lobes of the nucleus (a measure useful in determining the age of a white blood cell).

The final two rows of Figure 9 relate to the one-bit nuclear image which is created by thresholding at the average nuclear transmission level (see row 8). In the final row, a labeling operation has been applied to measure nuclear fine structure. A count of the pixels labeled provides a measure of nuclear texture which has been found useful in distinguishing small monocytes from large lymphocytes [27].

The entire sequence used for image analysis in diff3 is about 500 Golay Transform steps in length. The image processor in diff3 requires about 300 milliseconds to execute this sequence over the 4096 pixel field employed. Since each Golay Transform is equivalent to at least 20 assembly language instructions in a general-purpose computer, the image processor in diff3 is at least equivalent to a 100 MIP (million instructions per second) machine.

TISSUE-BIOPSY IMAGE ANALYSIS

The examination of the tissue biopsy [28] is as frequently performed in biomedicine as is the examination of the blood smear described above. Visual tissue biopsy examination is done in the hospital pathology laboratory. Unlike blood smear analysis, where there is a great deal of uniformity from input to input, the work of the pathologist is characterized not only by the variety of tissues examined but also by the complexity of the analysis required. The task of the pathologist is comparable in difficulty to that performed by the radiologist in examining the x-ray. The section reports preliminary research in applying neighborhood logic to digitized images of diseased liver tissue biopsies for the purpose of computer localization of disease sites.

Liver Tissue Architecture

Tissue biopsies are prepared by sectioning tissue with a microtome which produces a thin (a few micrometers) slice of tissue which is mounted on a microscope slide. As with the blood smear, the tissue slice is stained with certain biochemicals in order to provide information

Figure 8.   Stained buffy coat smear of human blood showing the standard Wright/Giemsa stain. (Courtesy Dr. L. Ornstein, Mt. Sinai Hospital, New York.)

Figure 10.   Image produced by the Automatic Light Microscope System (Jet Propulsion Laboratory, Pasadena, CA, U.S.A.) of a 400 × 400 μm region of a slice of fatty human liver tissue.

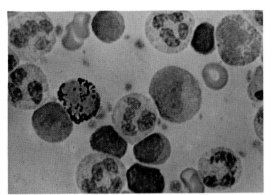

Figure 8

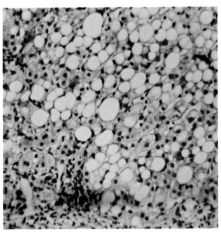

Figure 10

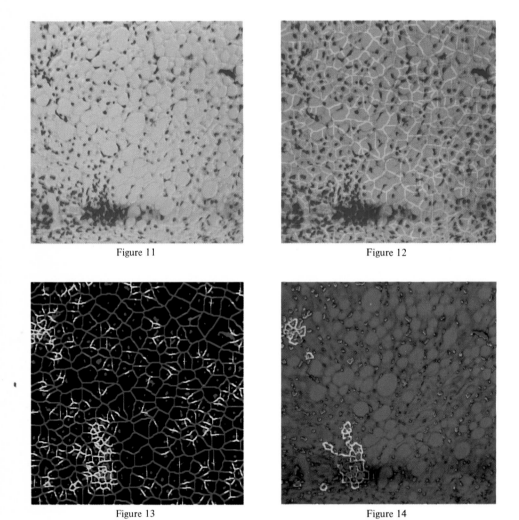

Figure 11

Figure 12

Figure 13

Figure 14

Figure 11. By thresholding the green component of Fig. 10 at the mid-range of its probability distribution function it is possible to locate cell nuclei.

Figure 12. The complete exoskeleton of the cell nuclei located in Fig. 11 is shown superimposed on both the cell nuclei and the red component of the original image.

Figure 13. Growth of the exoskeleton showing filaments which are formed early (white), those which are formed at an intermediate stage (blue), and those which are formed during the final stages (green).

Figure 14. Closed loops in the partial exoskeleton are generated from the white filaments shown in Figure 13. These closed loops are found to be specific to disease sites in the tissue.

on the chemical composition of tissue constituents (cell nuclei, cyto-
plasmic granules, connective tissue structure, etc.) by colorimetric
labeling. The tissue itself may be drawn from an excized organ (or
portion thereof) or from the percutaneous extraction of a tissue core
with a biopsy needle. The biopsy needle is hollow and the core which
it extracts is approximately 10 mm long and 1 mm in diameter.

Figure 10 is a full-color image of a portion (400 × 400 μm) of
a slice of a needle biopsy core taken from a fatty human liver. The
white regions consist of the cytoplasm of cells that have metabolised
fat, the small black objects are the nuclei of the tissue cells. Many
types of cells are present, e.g., ordinary liver cells (hepatocytes),
various types of white blood cells (leukocytes), red blood cells, plasma
cells, epithelial cells, fibroblasts, kupfer cells, etc. Cell-by-cell identification is difficult even for the trained observer. Major disagreements
occur from pathologist to pathologist on certain cell types. In some
cases, no identification is possible. Rather, the pathologist ordinarily
diagnoses the state of the tissue from its overall architecture. In Figure
10 abnormalities are some fibrosis (lower right), some inflammation
(upper left), and in the lower left, a severe infection which is indicated
by an infiltration (concentration) of leukocytes. The direction taken in
the investigation which is reported here was to devise sequences of
neighborhood operations which would extract regions of abnormality.
The result was the discovery that certain features of the exoskeleton of
the cell nuclei (or clusters of cell nuclei) were the most useful
indicators of abnormal tissue architecture.

Method of Approach

The image analyzing logic used was based on generating a one-bit
image by thresholding the green component of Figure 10 at the mid-
range of its normalized probability-distribution function (Figure 11).
This one-bit image was then modified by examining each pixel and its
eight neighbors. The count of the total number of one-valued neighbors
was compared with a threshold (FAC) and the crossing number [29] of
the binary neighborhood string was compared with a second threshold
(CNUM). If the total count of one-valued neighbors exceeded FAC or
if the crossing number was equal to or greater than the threshold
CNUM and the value of the pixel examined was a binary one, then the
corresponding pixel in the output image was made equal to a binary
one.

Many neighborhood logic sequences were investigated. The structure of these sequences and their outcomes were examined and it was found that the most useful sequence for locating abnormalities in the tissue structure was that which found closed loops in the partial exoskeleton. The exoskeleton (Figure 12) was generated by iterating using FAC = 6 and CNUM = 4 until there were no further changes in the output image. It was noted that, as the exoskeleton was generated, closed loops in the exoskeleton first appeared in regions of abnormality (Figure 13). By arresting the generation of the exoskeleton (i.e., the partial exoskeleton) after four iterations closed loops were extracted after 128 additional iterations with FAC = 8 and CMUM = 4. This relatively lengthy sequence culminated in the result shown in Figure 14 which was judged to be the most successful sequence for extracting regions of abnormality from the particular image under examination. It is now planned to evaluate this image processing method as a technique for automating certain determinations in histopathology.

ACKNOWLEDGEMENTS

The author wishes to thank Ms. C. A. Haluska, Department of Electrical Engineering, Carnegie-Mellon University, for typing the manuscript and Mr. G. B. Arnold and his staff for preparing Figure 1 and Table 1. The Dicomed Corp. (Minneapolis, Minnesota, USA) is acknowledged for generating Figure 10 through 14 from magnetic tapes furnished by the author. The original tissue image (Figure 10) was produced by the ALMS (Automatic Light Microscope Scanner) at Jet Propulsion Laboratory (Pasadena, California, USA) under a grant from the United States National Science Foundation. Other contributors are acknowledged in the figure captions. Finally, this work would have been impossible without the assistance of Dr. N. Wald and his staff (Drs. J. M. Herron, L. Davis, and S. England), Department of Radiation Health, University of Pittsburgh, who furnished time on their Biomedical Image Processing Unit as well as the many hours spent by Dr. A. Dekker of the School of Medicine in consulting on liver pathology.

REFERENCES

1. United States Dept. of Commerce, 'Statistical Abstracts of the United States', 1975.

2. E. H. Land *et al.,* 'A Color Translating Ultraviolet Microscope', Science 109, 371-374, 1949.

3. V. K. Zworykin and V. K. Hatke, 'Ultraviolet Color-Translating Microscope', Science 126, 805-810, 1957.

4. W. H. Walton, 'Automatic Counting of Microscopic Particles', Nature 169, 518-520, 1952.

5. D. Causley and J. Z. Young, 'Flying Spot Microscope', Science 109, 371-374, 1955.

6. K. Preston, Jr., 'The CELLSCAN System: A Leukocyte Pattern Analyzer', Proc. Western Joint Comput. Conf., p. 173, 1961.

7. N. F. Izzo and N. Coles, 'Blood Cell Scanner Identifies Rare Cells', Electronics 35, 52-57, April 27, 1962.

8. M. Ingram and K. Preston, Jr., 'Automatic Analysis of Blood Cells', Sci. Amer. 223(5), 72-82, 1970.

9. G. Lodwick *et al.,* 'The Coding of the Roentgen Image for Computer Analysis as Applied to Lung Cancer', Radiology 81, 185-200, 1963.

10. M. L. Mendelsohn *et al.,* 'Computer-Oriented Analysis of Human Chromosomes - I. Photometric Estimates of DNA Content', Radiology 81, 185-200, 1966.

11. G. L. Wied *et al.,* 'Taxonomic Intra-Cellular Analytic System (TICAS) for Cell Identification', Acta Cytol. 12(3), 180-204, 1968.

12. G. N. Hounsfield, 'A Method and Apparatus for Examination of a Body by Radiation Such as X or Gamma Radiation', Brit. Patent Spec. 1283915, London, 1972.

13. D. A. Cotter and B. H. Sage, 'Performance of the LARC tm Classifier in Clinical Laboratories', J. Histochem 24(1), 202-210, 1976.

14. M. Levine, 'Automated Differentials: Geometric Data's HEMTRACK [tm]', Am. J. Med. Technol. 40, 462, 1974.

15. J. E. Green, 'Parallel Processing in a Pattern Recognition Based Image Processing System: The Abbott ADC-500[tm]Differential Counter', Proc. 1978 IEEE Comput. Soc. Conf. Pat. Recog. Im. Proc., p. 492-498, 1978.

16. G. I. Hines, 'Instrumentation in Nuclear Medicine', Volume 1, Academic Press, New York, 1967.

17. D. H. Howry, 'Techniques in Ultrasound Visualization of Soft Tissues', in Ultrasound in Biology and Medicine (E. Kelly, ed.), Waverly Press, Baltimore, p. 49, 1957.

18. A. M. Cormack, 'Reconstruction of Densities from their Projections with Applications to Radiological Physics', Phys. Med. Biol. 18(2), 195-207, 1973.

19. A. M. Cormack, 'Reconstruction of a Function by its Line Integrals with Some Radiological Applications', J. Appl. Phys. 34, 2722-2727, 1963.

20. D. E. Kuhl and R. Q. Edwards, 'Image Separation Isotope Scanning', Radiology 80, 653-661, 1963.

21. J. M. S. Prewitt, 'New Vistas in Medical Reconstruction Imagery', in Digital Processing of Biomedical Images (K. Preston, Jr. and M. Onoe, eds.), Plenum Press, New York, p. 133, 1976.

22. R. A. Robb, L. D. Harris, P. A. Chevalier, and E. L. Ritman, 'Quantitative Dynamic Three-Dimensional Imaging of the Heart and Lungs by Computerized Synchronous Cylindrical Scanning Reconstruction Tomography', in Roentgen-Video Techniques for Dynamic Studies of Structure and Function (P. Heintzen, ed.) G. Thieme, Stuttgart, p. 285, 1978.

23. B. K. Gilbert, M. T. Storma, C. E. James, L. W. Hobrock, E. S. Yang, K. C. Ballard, and E. H. Wood, 'A Real-Time Hardware System for Digital Processing of Wide-Band Video Images', IEEE Trans. Comput. C-25(11), 1089-1100, 1976.

24. M. J. E. Golay, 'Hexagonal Parallel Pattern Transformations', IEEE Trans. Comput. C-18, 733-740, 1969.

25. K. Preston, Jr., 'Feature Extraction by Golay Hexagonal Pattern Transforms', IEEE Trans. Comput. C-20, 1007-1013, 1971.

26. K. Preston, Jr., 'Use of the Golay Logic Processor in Pattern Recognition Studies Using Hexagonal Neighborhood Logic', in Computers and Automata, Polytechnic Press, New York, p. 609, 1972.

27. K. Preston, Jr. and J. R. Carvalko, 'On Determining Optimum Simple Golay Marking Transformations for Binary Image Processing', IEEE Trans. Electronic Comput. C-21, 1430, 1972.

28. K. Preston, Jr., 'Digital Picture Analysis in Cytology', in Digital Picture Analysis (A. Rosenfeld, ed.), (Springer Verlag, Heidelberg), p. 209, 1976.

29. E. S. Deutsch, 'On Some Preprocessing Techniques for Character Recognition', in Computer Processing in Communications, Polytechnic Press, New York, p. 221, 1969.

THE PROCESSING OF X-RAY IMAGE SEQUENCES

K. H. Hoehne, M. Boehm, G. C. Nicolae
Institut fur Mathematik und Datenverarbeitung
in der Medizin
Universitaetskrankenhaus Hamburg-Eppendorf und
Deutsches Elektronensynchrotron (DESY)
Hamburg
Federal Republic of Germany

1. INTRODUCTION

Among the variety of image-processing applications, the analysis of x-ray data is of growing importance. In the past, study efforts concentrated mostly on the analysis of static pictures. Today, there is an increasing interest on the part of the physicians in extracting functional parameters from time-sequenced x-ray pictures (angiography). An angiographic analysis (e.g., of the kidney) proceeds as follows (see Fig. 1): The physician applies a radio-opaque contrast medium to the kidney to be examined. The contrast medium propagates through the kidney vessels at the speed of blood circulation. The resulting scene, which is of constant morphology but varying intensity, is viewed by the image intensifier video system of the x-ray equipment. Usually, the scene is assessed by estimating differences from image to image to get qualitative information about blood dynamics. Potentially, however, the sequence contains spatially very differentiated quantitative information about blood velocity or the filtration function of the kidney, which cannot be extracted in sequential viewing by eye. Thus, it is desirable to use image-processing tools to extract these parameters.

Before investigating this problem, it is useful to look at the present situation of image processing in radiology. The ideal imaging technique in radiology would be one that delivers the x-ray absorption

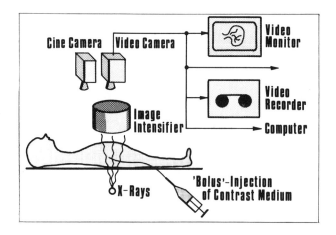

Figure 1.   Principle of angiography.

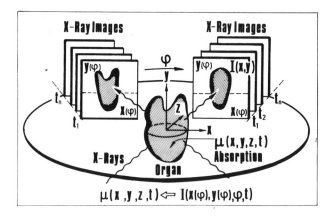

Figure 2.   Schematic diagram of image formation in radiology.

of the examined organ as a function of space and time. As illustrated in Fig. 2, this could, in principle, be achieved by taking image sequences of the organ from several (typically 180) directions. Although there are experimental setups proposed which allow the acquisition of such complete information [1], the involved amounts and rates of data are prohibitive to clinical application in the near future. Consequently, we have to confine ourselves to subsets (see Fig. 3) of data which are specified on one hand by technology and on the other hand by the problem to be solved. The most simple subset is the conventional projection *radiograph*. The image-formation process here is a single projection. Image-processing procedures have concentrated on automatic detection of size, shape and texture, and classification of these properties into diagnostic categories. Another procedure is that of *computer tomography,* producing maps of absorption values parallel to the x-ray direction of a set of projections. The resulting images are assessed by the physician directly or, in some cases, processed in a way similar to what is done with conventional radiographs. Comparing these two main image-processing techniques, we find that the processing of conventional radiographs has no clinical relevance yet, whereas computer tomography is in wide-spread use. The reason for this is obvious. In the first case, the image-*interpretation* process has to be automated by appropriate algorithms. This has proven to present very complex problems. In the second case, the *image formation* is done by the computer, converting a noninterpretable data structure (the projections) into a pictorial structure (the computer tomogram) which the physician may interpret as he is accustomed to. Although there are implementation problems, the algorithms for reconstruction are comparatively simple. Since, in our special application, we are interested in the time course of a process, we consider the subset of x-ray data containing one conventional projection as a function of *time.* Having in mind that improved images, or images containing new information, are more helpful to the physician than automatic interpretation, our investigation focussed on the development of algorithms for image formation rather than image interpretation.

Earlier, approaches to the problem of the quantitative measurement of blood dynamics from an x-ray image sequence used the technique of angiodensitometry [2]-[8]. In the case of nonmoving organs (e.g., the kidney) the time course of x-ray intensity at one or several regions selected by the physician is registered. The analysis of the resulting curve yields parameters such as blood-stream velocity. For moving organs (e.g., the heart), the boundaries of the region of interest are determined first; then the dynamic behavior of the parameter describing the region, such as the area of a ventricle, is computed. Even in the simpler case of nonmoving organs, the angiodensitometry is far from clinical application for the following reasons:

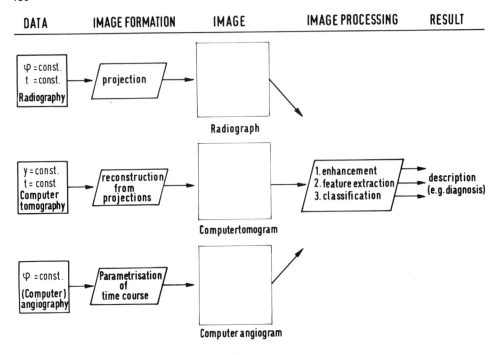

Figure 3. Different scopes of image processing in radiology.

Unless advanced image-processing tools are used, the analysis procedure is too laborious for utilization in a clinical environment.

If, nevertheless, a parameter such the blood velocity in a single
vessel is computed, one recognizes that there is no unique reference
value with which it could be compared because of the complexity of
the kidney function.

Therefore, if only relative differences for various regions of the same
organ are measured, one has the problem of comparing a large set
of numbers which, especially at the high resolution of x-ray images,
is prohibitive to practical application.

Our approach to the problem of analysis of an x-ray-image
sequence is based on the fact that an appropriate pictorial presentation
of the parameters derived from the sequence is best suited to combine

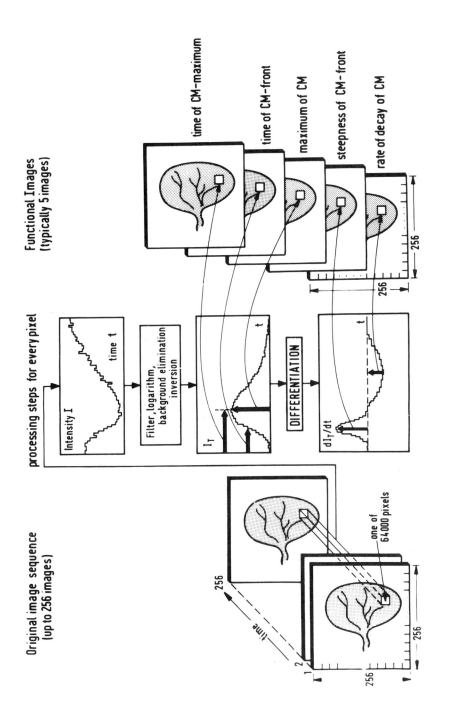

Figure 4. Principle of computer angiography.

the ability of the computer to do computations on large amounts of data and the ability of man to interpret complicated pictorial structures. The method, named 'Computer Angiography' [9], [10], is described in the following chapter.

2. COMPUTER ANGIOGRAPHY

2.1 Image Coding

As the picture sequence, in general, is highly redundant, it must be compressed to have a manageable amount of data. Typically, it consists of 128 frames of 256 × 256 pixels with eight-bit gray-level resolution. It is now encoded in a way which is based on the fact that only the characteristics of the intensity variation with time contains the information the physician is interested in. Thus the following procedure is applied (see Fig. 4): At any pixel, the variation of intensity with time caused by the wash-in and wash-out behavior of the contrast medium (CM) is considered. Its shape, in the case of the kidney, a curve with a steep rise, a variable plateau and a slow decay, is now described in a first-order approximation by featuring data such as the *time of the CM maximum, time of the CM front, maximum of CM, steepness of the CM front* and *rate of CM decay.* Instead of storing and processing the full-length curve (128 bytes), we now have to process a vector of five bytes only for each pixel which is a manageable size for routine application. It will be shown in a later section that this coding scheme preserves at least the information the physician draws from the sequential viewing.

2.2 Visualization of Blood Dynamics by Functional Images

The decisive point in our approach to extract more information than by sequential viewing is the following: Instead of presenting the physician a sequence of pictures as a function of time, we give him a set of static pictures, each of which describes one point of view of the dynamic behavior of the organ. This is achieved simply by presenting the feature matrices as gray tone or color pictures. Thus, the image *time of the CM front* is representative for the blood velocity, whereas the image *maximum of CM* is representative for the blood perfusion. Figure 5 shows a functional image of the type *time of the CM front,* together with one frame of a conventional angiogram.

We recognize the following properties:

Formally, it is very similar to the conventional one. Thus, the radiologist does not have to change his practice of interpretation.

The *semantics,* however, is completely different. It shows a survey of the blood speed during the observation time. In other words, it transforms functional information into static *morphological* information which can be assessed by the physician in the way he is used to.

As a decisive advantage, any background from overlaying tissue or device inaccuracies disappears.

As shown in Fig. 9, a quantitative analysis is possible by presenting the image in color code. Nevertheless, here too, the radiologist assesses the image from its morphological appearance.

Functional imaging is also being used in nuclear medicine [10]-[13] but has not reached clinical applicability yet. This seems to be due to inferior spatial and time resolution.

2.3 Computation of Global Function Parameters

Functional images present *local* functional parameters dependent on the morphology of the organ. From the data, global parameters which do not depend on the morphology can also be derived. Figure 6 shows, as an example, the frequency histogram of the parameter time of the CM front. From the time of the decay of the histogram, the transit time from the arteria to the cortex of the kidney can be computed in a simple way.

2.4 Computation of Functional Histograms

The elements of a functional image represent only one parameter. The information which is contained in the combination of parameters is not used. Thus, the optical presentation of two-dimensional histograms of parameter pairs is a tool for visualizing patterns which could contain more detailed information on the organ function. These histograms can be considered as some kind of a second-order functional image which, of course, no longer contains the morphological information the physician is used to. Suspicious regions, however, may be retransformed into the morphological picture. Figure 7 shows, as an example, the histogram time of CM front versus time of CM maximum for a kidney.

2.5 Classification According to Physiological Criteria

Should the investigation of the two-dimensional histograms, which is still done with the human eye, deliver promising results, the classification of the multidimensional parameter combination according to physiological criteria is to be considered. The projection of the classes (e.g., healthy or damaged tissue) into the morphological image may then lead to *refined functional images.*

3. IMPLEMENTATION

Except for classification, the algorithms are fairly simple. For a thorough test and a final utilization in *clinical environment,* however, there are constraints concerning the hardware and software *implementation.* Clinical environment means that the users are not at all interested in image-processing algorithms but need a quick availability of a medical result. Thus, hardware and software structures which are fast and hide complex internal structures from the medical user must be found. In the following chapter, we describe the implemented system we named CA-1 (Computer Angiography System-One).

3.1 System Hardware

Requirements. The requirements were that the system had to be fast, and the *aquisition* of the picture series had to be performed at a speed allowing immediate analysis. As a consequence, we chose the video signal as input and aimed at a real-time digitization of video information. The analysis also had to be fast because it was required to be carried out as an interactive process. A routine application of this technique is only possible if it is not time consuming. Furthermore, should the result of a processing step require the acquisition of new data, then it must be available during the time the patient is still present. To meet these requirements, one must decide between two approaches. A pure real-time processing of multitemporal images as proposed by Kruger *et al.* [15], [16], on one hand has the advantages that some simple algorithms are fast and no temporary storage of large amounts of data is necessary, which makes the system cheaper; on the other hand, it has the severe disadvantage that more sophisticated algorithms cannot be applied, and after the image sequence has passed, all information which could not be obtained in real time is lost. There-

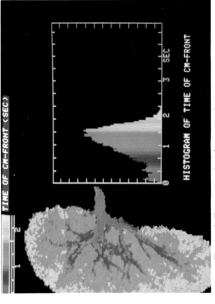

Figure 5. Functional image 'time of CM front' of a kidney, together with one digitized frame of the angiogram.

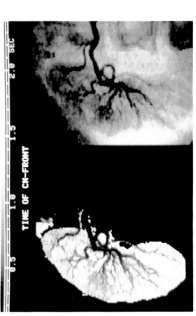

Figure 7. Example of a histogram for two parameters.

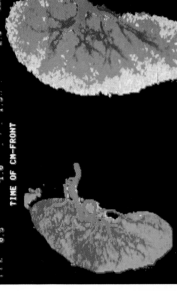

Figure 6. Example of a histogram for one parameter.

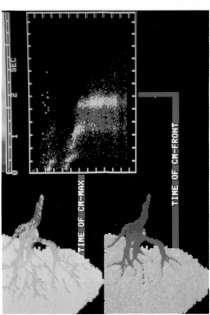

Figure 9. Functional images of a normal and an abnormal kidney.

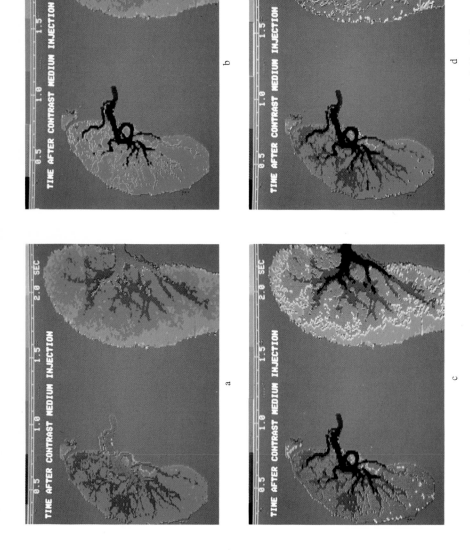

Figure 10. Four phases (Figs. 10a - d) of the real-time reconstruction of blood flow.

fore, we decided not to aim at pure real-time processing. Instead, we aimed at high-speed processing from stored data. One reason is that the time context necessary for our algorithms is too large. Another reason is that development and refinement of algorithms are much more difficult if they are implemented in hardware at a too-early-stage of research. In detail, the following functions have to be performed:

Picture-data acquisition, storage, display and transfer of pictures (or regions of pictures) at standard video rates (50 frames/s, 10 Msamples/s);

Picture-data processing such as the computation of functional images at a speed allowing interactive analysis;

Suitable communication between user and system by adequate means

of (graphics) command input; and

pictorial and graphic output which is fast enough for inter-active use and animation of pictures.

Concept. Considering that features required most urgently are flexibility and speed, we chose to realize a multiprocessor structure [14] called Digital Video System (DVS). Figure 8 shows the structure of this multiprocessor system. It consists of several dedicated micro-programmed processors which are connected by a high-speed asyn-chronous bus (60 ns cycle time). The high-speed bus dynamically switches data paths between pairs of processors and performs the process synchronization according to the preassigned priorities.

Image buffers. Two image buffers of 32 k × 16 bits to store video image data are used as common resources by all processors. The image sequence buffer presently being implemented is used for storage and retrieval of up to four million pixels (four Mbytes). The memory/control processor gives the user the possibility to access the memory on a picture by picture basis or orthogonally on a 'time-course-of-pixels' basis. This decisively reduces the data reorganization time for the analysis algorithms which generally work on the time course.

Real-time digitizer. The Real-Time Digitizer (RTD) performs real-time acquisition, digitizing and storage of video images. In its design, we aimed at maximum data reduction immediately at acquisition time. This is achieved by the programmability of the acquisition param-eters, such as region of interest, time and spatial resolution. Thus, only limited regions are digitized with no more spatial and time resolution than necessary. The RTD contains three classes of instructions:

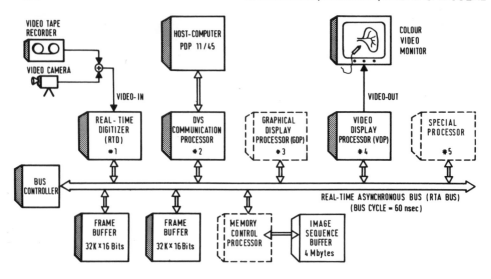

Figure 8.   Basic hardware structure of the system CA-1.

Instructions controlling the acquisition format. The user can select any rectangular window with $0 \leq X \leq 255$ and $0 \leq Y \leq 255$ for acquisition.

Instructions controlling spatial resolution (sampling rates of 5 MHz of 10 MHz) and time resolution (up to 50 frames/s).

Instructions controlling the destination of the digitized data.

Video-Display Processor. The Video-Display Processor (VDP) is used for the display of data contained in frame buffers or the color-video monitor. To give an easily-interpretable, optical presentation of the different semantics of the data, gray-level transformation and/or color coding is provided. Thus, the main feature of the VDP is its programmable real-time processing facility, based on the technique of fast look-up tables. Monadic and dyadic operations are available. Monadic operations work on one of the frame buffers or the concatenation of both. A typical monadic operation is a contrast enhancement. A dyadic operation uses the two frame buffers as operands. A typical application is image subtraction. As the look-up tables can be updated up to 50 times/s, a dynamic interpretation of the picture data is possible which even allows the reconstruction of moving scenes of images containing time parameters. The VDP contains three classes of instructions:

> Instructions controlling the display format meaning that any rectangular window with $0 \leq X \leq 255$ and $0 \leq Y \leq 255$ can be selected by the user.

> Instructions controlling the spatial resolution (256 or 512 pixels/line).

> Instructions for filling the look-up tables (256 words, eight bits, and 4096 words, nine bits).

> Instructions controlling the data paths.

Video Tape Controller. For the long-term storage of image sequences, a Video-Tape Controller was built which, together with the Real-Time Digitizer, makes the video tape look like a digital tape from which any image is addressable. Thus, about 350 scenes of 10 seconds each may be stored and retrieved.

The structure of the DVS is such that it is open to new processors. For instance, algorithms tested in the host computer may be implemented in a special processor for fast execution. The system has now been working successfully for two years.

3.2 System Software

The main requirements to be met by software implementation are, on one hand, high efficiency, necessary to process the large amounts of multi-temporal image data in acceptable time, and, on the other hand, enough flexibility to use the system in an interactive way.

Concept. To achieve these features, we implemented an experimental dialog system called 'PROFI-11' (Processing and Retrieval of Functional Images on a PDP-11 Computer). To achieve flexibility, PROFI-11 includes dialog features like immediate execution of single commands and programs which have been edited beforehand. User-definable function keys are provided for frequently-used commands or programs. Programs may be interrupted by user interruption, e.g., to enter commands for inspection or modification of variables. Afterwards, the program execution may be continued. Efficiency was achieved by using the concept of a simple table-driven compiler instead of a time-consuming interpreter and an implementation language (SIMPL-11) previously designed [17], which combines efficiency and machine orientation of assembly languages with the advantages of control structures and data types as known from high-level languages.

The system is partitioned into a frame (supervisor, table-driven compiler, data-definition facilities) and a set of problem-oriented software processors. These processors, the largest part of the system, use the standard FORTRAN subroutine interface, so that they may used without the frame as libraries for stand-alone programs, or be incorporated into other frames, e.g., a dialog language for image processing [18] implemented concurrently. This language will also support interested but untrained physicians.

Data Definition and Data Types. The data-definition facilities include, besides basic data types INTEGER and VECTOR, the problem-oriented data types COLOR, IMAGE, and MULTIPLE IMAGE. To define the total size or to access subsets of images and multiple images, we use the data types REGION (one-dimensional), FRAME (two-dimensional) and CUBE (three-dimensional). For better legibility, the names of such objects are preceded by special signs (% for regions, # for frames, and @ for cubes). Typically, the regions %X and %Y are defined by

REGION %X, 100, 128;
REGION %Y, 50, 200.

A frame named XY may be composed of both regions by

FRAME XY, %X, and %Y.

Then, an image located physically in image buffer 1 (indicated by M1) may be defined by one of the following equivalent commands:

IMAGE TSTIMA, #XY, M1;
IMAGE TSTIMA, %X, %Y, M1; or
IMAGE TSTIMA, 100, 128, 50, 200, M1.

During a session, new data objects may be defined, old ones inspected via ASK-command or dropped via DROP-command. For suitable objects (e.g., frames) an interactive definition or alteration by trackball is also provided.

Operators. A large set of monadic and dyadic operators (including byte and word arithmetic with and without automatic scaling, bit-handling, I/O operators) is implemented only once in a common operator pool shared by the operations on integers, vectors, images and multiple images. The execution speed of complex monadic byte operators is greatly increased by using the concept of computing a look-up table only once for the operator, afterwards used as a fast transformation map (e.g., for logarithmization of images). For image enhancement and analysis, we use iterative neighborhood techniques having important advantages over pure local or sequential approaches. Pure local approaches are too noise sensitive and the results of sequential approaches will often depend upon the order in which the pixels are examined. All neighborhood algorithms are implemented using a common program frame for effective fetching of operands and storage of results.

Specific Hardware Support. To achieve flexibility, it is necessary that specific hardware requirements and restrictions are hidden from the user. For example, the user may work with images located physically in one of the frame buffers in the same way as images located in memory. Also, it is not necessary to know how to fill the hardware look-up tables to perform a specific monadic or dyadic image operation. The corresponding look-up table may be derived from any byte operator automatically.

Manipulation (shifting and scaling) of tables for gray-level transformation or pseudo-color coding may be done interactively by track-ball.

Database Facilities. To provide local database facilities, a simple, relational database system was implemented. It is based on the file system, allows definition of new relations and offers the necessary selection, deletion, addition and update features. As the involved data amounts are small, all searches are purely sequential. For long-term storage, the local database will be connected to a large relational database system [20] implemented in parallel on an IBM 370/168 on top of the commercial linear database system ADABAS.

Performance. PROFI-11 is implemented on a PDP-11/45 computer. For our application, the average processing speed was doubled by adding a cache-memory. The following examples will give an impression of the system's execution times:

Monadic image operations: To clear an image ($256 \times 256 \times 8$ bits) takes 0.25 seconds. Logarithmization of such an image with explicit computation of the logarithm for each pixel takes 45 seconds. Using a software look-up table, the execution time is only 1.7 seconds.
Dyadic and neighborhood operations: Adding or subtracting two images ($256 \times 256 \times 8$ bits) takes 2.0 seconds. A sample 3 $\times$

3-smoothing algorithm takes 15 seconds. A more complex noise-cleaning algorithm which considers a 3 × 3 neighborhood will probably belong to the same region (therefore sorting of neighborhood is necessary) takes 21 seconds to execute.

Multitemporal images: To compute a set of four functional images from a series of 128 images of an organ fitting into a 128 × 128 picture matrix takes four minutes. Most of this time is necessary for noise cleaning of intensity versus time curves. We are presently investigating whether parameters can be found with comparable physiological meaning but less noise sensitivity.

4. APPLICATION

With the described method, about 140 image sequences (120 kidneys and 20 livers) have now been examined. The following examples are to show that, already in the present state, results with clinical relevance have been achieved.

Figure 9 shows the functional images of the type 'time of CM front' for a normal kidney (left) and a kidney with a nephrotic syndrome (right). At first glance, the absence of device-dependent intensity inhomogenities and overlays is obvious, which leads to a clear presentation of the organ. The radiologist can immediately infer from the color distribution that the blood flow in the right kidney is delayed. By comparison with the color scale, he can even quantify the delay. This diagnostically relevant information is not obtainable from the conventional angiogram. As these functional images represent 'time tables' for contrast medium propagation, the original sequence may be reconstructed from them and played back by the CA-1 system in real time (see Fig. 10). The visual impression is nearly the same as from the original sequence. We, therefore, conclude that the main part of the information contained in the sequence is already represented in one single functional image. The relevance of the method for therapy control is illustrated in Fig. 11 which shows the functional image 'time of CM front' for a kidney before (Fig. 11a) and after (Fig. 11b) the injection of Angiotensin, which produces a contraction of the blood vessels. By comparison of the images with the fiducial scale, one easily obtains the result that the velocity of the blood stream was reduced by 40 % to 50 %. A study of this subject is being published [19].

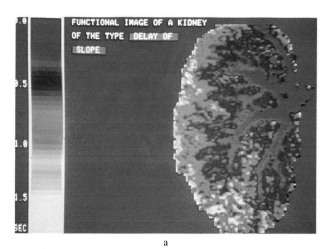

a

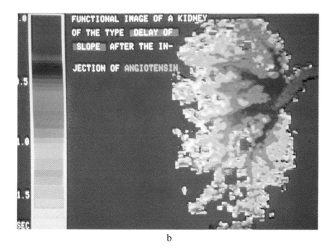

b

Figure 11. Functional images 'time of CM front' for a kidney before
(Fig. 11a) and after the injection of angiotensin (Fig. 11b).

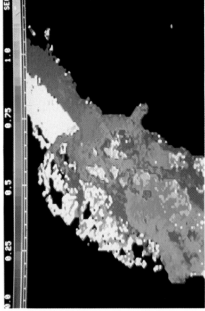

Figure 12. Functional image of a transplanted kidney.

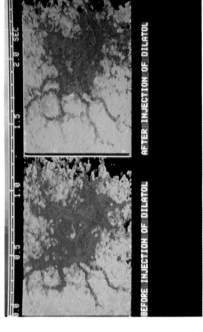

Figure 13. Functional image 'time of CM maximum' for a kidney with a tumor.

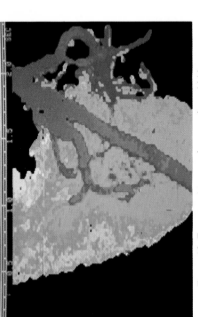

Figure 14. X-ray intensity as a function of time for different picture elements of a kidney.

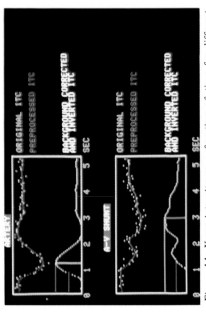

Figure 15. Functional images of a liver before (left) and after (right) treatment.

A very promising application started recently is the control of transplanted kidneys. Here it is very important to discover the symptoms of a rejection at a very early stage to apply the appropriate therapy. Figure 12 shows a functional image of the type 'time of CM front' of a transplanted kidney. In this case, the delayed blood perfusion in the upper pole (yellow-brown region) could be recognized earlier than by conventional methods. Figure 13 is to demonstrate that even more complicated defects in the blood dynamics may be localized by Computer Angiography. It shows a functional image 'time of CM maximum' of a kidney with a tumor covering nearly the whole organ. The yellow region in the center gives the impression that there is a region with delayed blood flow. The analysis of the x-ray intensity as a function of time in the suspicious region gives the result that this region describes a shunt between the arterial and the venous system. In this case, the curve (Fig. 14) shows a twin peak corresponding to the arterial and the venous flows, respectively. The program, accidentally, interprets this situation as a delayed arterial flow. We conclude that the refinement of the algorithms to recognize more special features such as the presence of twin peaks may lead to the localization of abnormalities not visible at all in conventional angiograms.

The method is, of course, applicable to other organs such as the liver, the brain and the heart. Figure 15 shows the functional image 'time of CM front' of a liver before (left) and after the injection of Dilatol which is known to increase the blood flow (right). In the case of this patient, who was suffering from a liver cirrhosis, it could be proven by the functional image that there was no significant influence on the blood flow.

5. CONCLUSIONS

It was shown that the new method of Computer Angiography is an instrument for obtaining essentially more and better diagnostic information from angiograms of organs such as the kidney and the liver. Although the radiologist gets a new kind of information, he does not have to change his technique to interpret images. The proven success of the described method supports our opinion that, at the present state of medical image processing, one most promising approach is to compute new images containing better information rather than aim at the automization of image interpretation, which is still done best by the radiologist.

REFERENCES

1. S. A. Johnson *et al.*, 'Bioimage Synthesis and Analysis from X-Ray, Gamma, Optical and Ultrasound Energy' in *Digital Processing of Biomedical Images* by K. Preston and M. Onoe, Eds., Plenum Press, New York, 1976, 203.

2. N. R. Silverman, 'Videometry of Blood Vessels', Radiology 101 (1971), 697.

3. N. R. Silverman, M. Intaglietta, and W. R. Tompkins, 'Renal Cortical Transit Times', Radiology 105 (1972), 25.

4. L. Rosen and N. R. Silverman, 'Application of Videodensitometry to Quantitative Radiological Measurement in Medicine', Proc. of the Soc. of Photo Optical Instr. 40 (1973), 69.

5. K. Vanselow and F. Heuck, 'Neue Grundlagen und Theorien zur Verbesserung der Angio-Cine-Densitometrie', Fortschr. Roentgenstr. 122 (1975), 453.

6. E. Epple, D. Decker, F. Heuck, 'Mess- und Auswertverfahren der Roentgen-Cine-Densitometrie mit einer Datenverarbeitungsanlage', Fortschr. Roentgenstr. 120 (1974), 345.

7. K. H. Deininger, 'Angiographisch densitometrische Moeglichkeiten zur Bestimmung der renalen Blutdynamik', Med. Welt 27 (1976), 1764.

8. H. D. Pieroth, E. Magin, M. Georgi, and M. Marberger, 'Ergebnisse der kineangiodensitometrischen Nierendurchblutungs-bestimmung', Fortschr. Roentgenstr. 126 (1977), 126.

9. K. H. Hoehne, M. Boehm, W. Erbe, G. C. Nicolae, G. Pfeiffer, and B. Sonne, 'Computer-Angiography - A New Tool for X-Ray Functional Diagnostics', Med. Progr. Technol. 6 (1978), 23.

10. K. H. Hoehne, M. Boehm, W. Erbe, G. C. Nicolae, G. Pfeiffer, B. Sonne, and E. Buecheler, 'Die Messung und differenzierte bild-liche Darstellung der Nierendurchblutung mit der Computer-Angiographie', Fortschr. Roentgenstr. 129 (1978), 667.

11. S. Kaihara, C. Natarajan, D. Maynard *et al.*, 'Construction of a Functional Image from Spatially Localized Rate Constants Obtained from Serial Camera and Rectilinear Scanner Data', Radiology 93 (1969), 1345.

12. H. Agress, Jr., M. V. Green, and D. R. Redwood, 'Functional
     Imaging Methodology: Recent Clinical and Research Applica-
     tions', Proc. of the IVth Intern. Conf. on Information Processing
     in Scintigraphy (1975), 1289.

13. H. Toyama, M. Ilio, J. Iisaka *et al.*, 'Color Functional Images of
     the Cerebral Blood Flow', J. Nucl. Med. 17 (1976), 953.

14. G. C. Nicolae and K. H. Hoehne, 'A Multiprocessor System for
     the Real-Time Digital Processing of Video Image Series', DESY-
     Report DV 78/2 (1978).

15. R. A. Kruger, S. J. Riederer, T. L. Houk, C. A. Mistretta, C. G.
     Shaw, and J. C. Lancaster, 'Real-Time Computerized Fluoroscopy
     and Radiography', A Progress Report (preprint).

16. R. A. Kruger *et al.*, 'Preliminary Studies of Real-Time
     Computerized Fluoroscopy Techniques for Noninvasive Visuali-
     zation of the Cardiovascular System' (preprint).

17. G. Pfeiffer, 'SIMPL-11 - Eine einfache Implementationssprache
     fuer PDP-11 Rechner', DESY-Report DV 76/2 (1976).

18. G. Pfeiffer and K. H. Hoehne, 'Improvement of Programming
     Efficiency in Medical Image Processing by a Dialog Language',
     Proc. of MIE '78, Lecture Notes in Medical Informatics, Vol. 1,
     Springer Publishing Co., New York, 1978.

19. W. Erbe, B. Sonne, and K. H. Hoehne, 'Eine quantitative
     Bestimmung der Wirkung von Angiotensin auf die Nierendurch-
     blutung', Fortschr. Roentgenstr. 130 (1979) (in press).

20. M. Boehm and K. H. Hoehne, 'Aufbau eines Datenbanksystems
     nach dem Relationenmodell fuer die medizinische Anwendung',
     Jahrestagung der GMDS, Goettingen, 1977 (in press).

LANDSAT IMAGE PROCESSING

E. E. Triendl
Deutsche Forschungs- und
Versuchsanstalt fuer Luft-
und Raumfahrt
Oberpfaffenhofen
Federal Republic of Germany

Over 15 years ago the first photographs of the earth
from space were taken with a hand-held camera on the
MERCURY 8 mission. LANDSAT 1, the first spacecraft designed specifically for earth observations, was launched
on July 23, 1972. Presently LANDSATs 2 and 3 are in
operation. LANDSAT D is scheduled for 1981 to possibly
form part of an operational earth observation system.
In addition to the LANDSAT program, SEASAT A transmits
high-resolution radar imagery. Large format cartographic cameras will be carried in the SPACE SHUTTLE and
EUROPEAN SPACELAB.

LANDSAT satellites have returned half a million
multispectral images or 10^{14} bit to earth. 500 papers
citing LANDSAT are published per year in areas such as
forestry, cartography, land-use, water monitoring,
geology, aquatic vegetation, evaluation of farmlands,
regional planning, highway engineering, census, pollutant
monitoring, bathymetry, snowcover forecasts, soil mapping,
crop inventory, wheat yield forecast, hydrology, forest
fire mapping, mosquito habitats, mineral resources, flood
monitoring, snowfield assessment, phosphate mining, ship
detection, air pollution, topographic mapping, change
detection, etc.

LANDSAT 1 which was built on the basis of the hardware of the successful NIMBUS G weather satellite and was
originally named ERTS-1 (Earth Resources Technology Satel-

lite) was officially retired in January 1978. An identical
LANDSAT 2 was launched in January 1975.

On March 5, 1978 a THOR DELTA rocket carried the improved LANDSAT 3 into a near polar, sun-synchronous orbit
from Vandenberg Air Force Base, California. LANDSAT 3 is
similar in design to the preceding LANDSAT; however, the
imaging sensors carried onboard have been modified. The
multispectral scanner (MSS) of LANDSAT 1 and 2 was designed to respond to earth reflected sunlight in four
spectral bands at wavelengths from 500 to 1.100 nano-
meters (green to near infrared) with an instantaneous
field of view of nominally 79 × 79 meters, whereas the
LANDSAT 3 MSS carries an additional band responding to
thermal infrared radiation, which will sense tempera-
tures from -13 ^{O}C to 67 ^{O}C with a temperature resolution
of 1.5 ^{O}C and a 237 m picture element. This band was meant
to also acquire nighttime thermal data during the satellites' ascending mode but the detectors failed recently.
The swath width of the MSS is 185 km. In place of the
three rarely used Return Beam Vidicon (RBV) cameras two
panchromatic RBVs produce side-by-side images which cover
99 × 99 kilometers with a 40 meter picture element.

LANDSAT D is presently targeted for launch in late
1981. It will be an experimental earth resources monitoring system incorporating a new seven-channel multispectral
scanner called the thematic mapper, together with a five-
channel multispectral scanner (MSS). The scanners will
be mounted on the Multi-Mission Modular Spacecraft (MMS)
which will be compatible with the shuttle retrieval and
replacement capabilities. It will be operated through
the geostationary Tracking and Data Relay Satellite System
(TDRSS) to produce an average of 50 TM and 200 MSS scenes
per day over selected regions of the earth. LANDSAT D
will be recovered and refurbished by the Shuttle. Together
with an identical LANDSAT D' it will possibly form an
operational Earth Resources Monitoring System in 1983.
Applications for the operational program will be:

   - crop acreage and yield assessment
   - water resources monitoring and management
   - mineral and petroleum exploration
   - land-use mapping and monitoring.

The system will be supplemented by large-format geographic
cartographic cameras and by national projects such as
SPOT (France) and JEOS (Japan). Since astronomers object
to their Large Space Telescope (LST) being misused for
earth observation, a Synchronous Earth Observation Satellite (SEOS) may be monitoring transient events in real
time in the mid-1980s.

THE MULTISPECTRAL SCANNER (MSS)

Since digital processing of Return Beam Vidicon (RBV) Data seems to be fairly uncommon, we deal only with the multispectral scanner (MSS).

Lightening conditions and coverage of a remote sensing satellite depend largely on the choice of it's orbital parameters. LANDSAT circles the earth at an altitude of 920 km in a sun-synchronous, near-polar orbit (inclination: 99 degrees). This orbit precesses about 1 degree per day, so that the spacecraft passes overhead at always the same local time (9:30 am at the equator and 10:12 am at latitude 60 degrees north). Adjacent ground tracks are spaced 159.38 km and covered on consecutive days, while it takes 18 days to cover the whole earth.

The LANDSAT Multispectral Scanner (MSS) is a line-scanning device that uses an oscillating mirror to scan perpendicular to the spacecraft velocity vector. Spacecraft motion provides along-track progression of the scan lines. Radiation is sensed simultaneously by an array of six detectors in each of the four spectral bands (a total of 24 detectors, LANDSAT 3 has two additional detectors for thermal radiation). The data are encoded and formatted into a continuous data stream of 15 mega-bits per second. The continuous strip imagery is trans-formed into framed images with 10 % overlap of consecutive frames and area coverage of 185 by 185 km by the Ground Data Handling System.

GEOMETRIC PROPERTIES OF MSS PICTURES

The raw MSS picture is distorted geometrically by overscan, earth rotation, changes in mirror velocity and the effects of earth curvature, map projection, the detector array and spacecraft manoeuvres. With the ex-ception of very rare spacecraft manoeuvres and the very small effect of the detector array, these distortions are basically quadratic in nature and can therefore be eliminated by a quadratic geometrical transform of the picture. The parameters of the transformation equation are usually obtained by the use of ground control points. These are points in the picture of which the location on the map is known precisely. The location of points taken from a library of ground control points may be de-termined either by hand or by correlation and optimi-zation techniques.

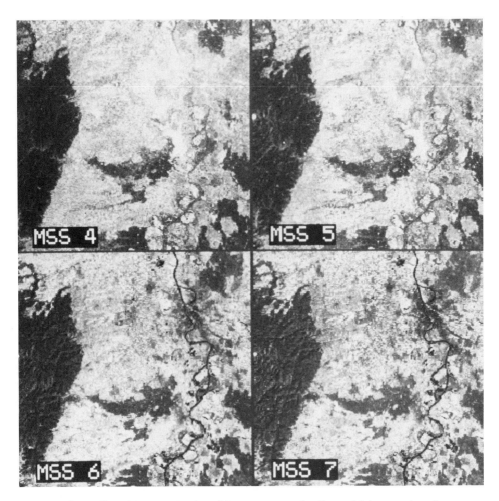

Fig. 1. Geometrically-corrected multispectral
scanner data. The scene corresponds to the 100 by 100 km
area of the map sheet "Mannheim" of the 1:200.000 scale
German map system. Since no single LANDSAT swath covers
this area, two adjacent pictures, taken at different
times, had to be merged.

After geometric transformation, the picture elements
have to be rearranged into a quadratic raster. This
process, known as resampling, can be done either by
using the intensity level of the nearest neighboring
picture element, or by bilinear interpolation of the
four nearest neighbors, or by cubical spline convolution
involving the surrounding four by four pixels.

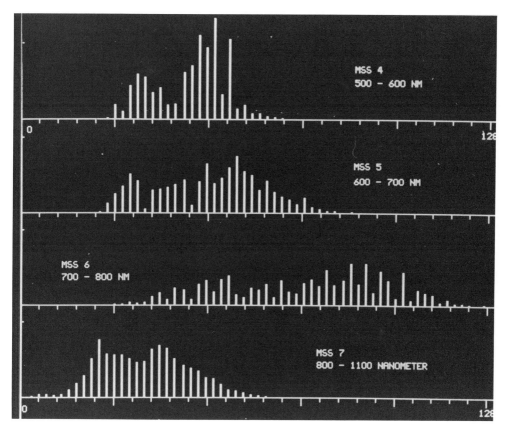

Fig. 2.   Gray-level histograms of the four channels

With geometric correction and appropriate radio-
metric adjustment several MSS-pictures can be mosaiced
together. Figure 1 shows a 100 by 100 km area, which
is covered by two adjacent swaths. Two scenes have been
geometrically corrected and merged together.

RADIOMETRIC PROPERTIES

The gray-level histogram (Figure 2) shows only a few
few dozen gray levels in each band. Moreover, the gray
levels of various bands are highly correlated. The
frequency of occurrence of pairs of gray levels is shown
in Figure 2 for the combinations MSS 4 / MSS 5, which
are highly correlated and for MSS 4 / MSS 7, which are
are the least correlated pair of bands.

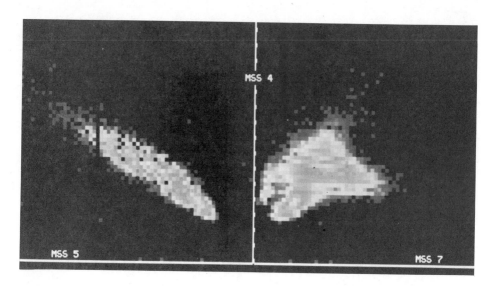

Fig. 3.   Two-dimensional histogram of MSS 4 / MSS 5
(highly correlated) and MSS 4 / MSS 7, the least correlated
pair of spectral bands.

The intensities of the four spectral bands span a four-
dimensional feature space. This space is populated only
within an area which has the form of a (three-dimensional)
slab. The principal component transform rotates this space,
so that its new base vectors - now called principal
components - point into the directions of maximum extension
of the population. Due to the slab shape of the populated
area components 3 and 4 are very small. For display the
principal components are contrast enhanced by normalization
of their variance. Figure 4 shows that components 3 and
4 carry in fact very little information (except noise).
The first principal component corresponds to brightness,
while the second may be attributed to color variations.
In other words: LANDSAT data may be considered to be
two-dimensional for many applications.

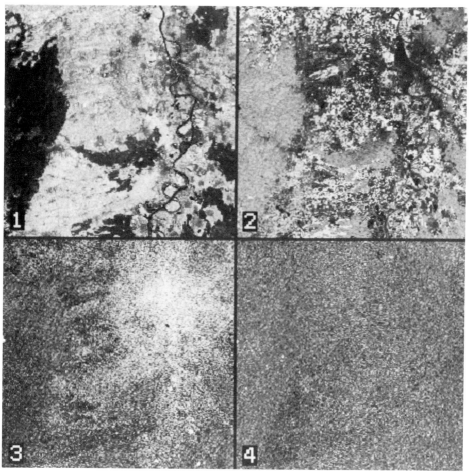

Fig. 4.   The contrast normalized principal components.
Components 1 and 2 carry useful information, while
components 3 and 4 contain mostly noise.

Pixel by pixel classification is performed by
assigning to each element of the feature space one out
of several classes, such as water, forest, town, etc.
The proceedure consists of two steps:
First the statistical class descriptions are derived from
areas with known land-use, so called training areas.
Then classification is performed for each picture
element.
There are many classification methods. A widely-used
one is the maximum-likelihood method. It approximates
the populations of the feature space by multi-variate
Gaussian distributions.

Fig. 6.   64 × 64 picture element clip of original
data corresponding to the center of the map in Figure 5.

Fig. 7.   Same as Figure 6, but with edges reconstructed.

Figure 5. Segment of a 1:200,000 scale land-use map combining LANDSAT data and cartographic features.

Legend:

blue:	water	brown:	city
light green:	gardens	red:	suburb/village
yellow:	fields	green:	deciduous forest
black:	unrecognizable	dark green:	pine forest

Geometric rectification plus classification produces a land-use map, which can be overlayed with other carto-graphic features. Figure 5 shows a segment of such a map. It was made in cooperation with the German Ministry for Construction, Organization and Regional Planning. This map cannot compete with the accuracy of land-use maps available in Germany, but is much more up to date and it is probably more accurate than maps available for many parts of the world. Its main problems are due to:

1) Pixel by pixel classification is not accurate enough. To get more accuracy either external data or multitemporal data are to be included. Multitemporal pictures are those taken at different times and registered upon each other, so that for each location not only radiation intensities in different spectral bands are available, but also for different times. These pictures are especially helpful for the classification of vegetation, which changes its color in the course of the year.

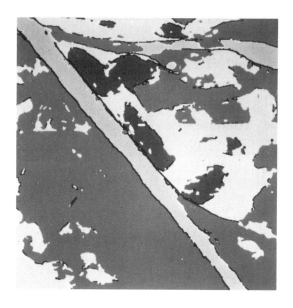

Fig. 8. Land-use mapping with reconstructed edges shows the gain in resolution possible with advanced methods (alas, too time consuming to be applied to the whole picture).

2) The seventy-meter picture element of LANDSAT is too
big for a 1:200.000 map. Especially when the classification algorithm makes use of neighborhood information —
either intrinsically or by subsequent application of a
cleaning algorithm — detail gets lost. This difficulty
may be overcome in part by model-directed interpretation.
Figures 6 to 8 give an example: It assumes that digitized
gray values are produced by step edges on the ground.
Where this assumption is proved true, the exact location
of the edge can be computed. In this way, superresolution
is obtained — in a sense. Figure 6 shows a 64 by 64 pixel
segment, the center of the map (Figure 5) with the Rhine
river flowing through the town of Mannheim.

From this reconstructed picture, a mapping at much
higher resolution can be produced. The experimental classification result (Figure 8) shows a much better deline-
ation of the river bed than pixel by pixel classification
is able to produce.

CONCLUSION

Geometric rectification and pixel by pixel classification of LANDSAT MSS imagery are fairly well understood and widely used. More advanced methods of picture
segmentation are very promising but still in the development stage.

Also better data will be available in the future,
i.e., more spectral bands, especially an infrared channel,
and higher resolution. Great use will also be made of
existing data bases such as maps, data from the same
satellite taken previously and data from other satellites
such as microwave and side-looking radar pictures. Every
single one of these new developments will also multiply
the computational requirements.

ACKNOWLEDGEMENT

All the material shown was made on DIBIAS (Digital
Interactive Bavarian Image Analysis System) by members
of the DIBIAS group at the Institute for Communication
Technology of the DFVLR. The map was made in cooperation
with the ministry for Construction, Organization and
Regional Planning.

SELECTED BIBLIOGRAPHY

LANDSAT data user handbook, NASA

Frederic J. Doyle, The Next Decade of Satellite Remote Sensing, Photogrammetric Engineering and Remote Sensing, February 1978

Ralph Bernstein, Digital Image Processing of Earth Observation Sensor Data, IBM Journal of Research and Development, January 1976

Ernst Triendl, How to Get the Edge into the Map, 4th International Joint Conference on Pattern Recognition, Kyoto, November 1978

IMAGE PROCESSING FOR DOCUMENT REPRODUCTION

P. Stucki
IBM Zurich Research Laboratory
8803 Rueschlikon
Switzerland

INTRODUCTION

The technical progress in almost every area of computer engineering is strongly related to cost and technology trends in micro-electronics and Input/Output (I/O) devices.

The steady refinement of Large Scale Integration (LSI) manufacturing techniques results in a constant price performance and package density improvement. The availability of lower-cost processors and semiconductor memories allows the transfer of subsystem functions from the host computer to decentralized locations. The evolution from centralized towards distributed computing will relieve host processors and transmission channels from executing low-function, high-volume data-processing tasks, as they typically occur when implementing document acquisition and presentation services. Consequently, host processing and data-transmission capacity become available for improved systems response.

Progress in the area of I/O devices can be traced direct to the advances in All-Point-Addressable (APA) scanner, display and printer technologies. Today, for example, an A4-size document can be scanned with a single linear charge-coupled device at a resolution of 200 picture elements (pel) per inch [1]. Color cathode-ray tubes are available that can display 1024 lines/frame, and ink-jet printers capable of depositing multi-color inks onto normal paper have made their appearance on the market [2].

With the logic power, the memory capacities and the new I/O devices available, the areas of impact on future information-processing systems and application areas are many.

THE AUTOMATED BUSINESS-ADMINISTRATION ENVIRONMENT

Today, digital computer systems are primarily used to collect, process, store and retrieve information in coded or alphanumeric form. Historically, the basic means of information reproduction has been the typewriter and the electromechanical printer which both provide a permanently recorded and legible hardcopy document. With the evolution towards higher system interaction and the aim to reduce the consumption of paper, the method of information reproduction had to meet new requirements. This trend led to the development of display terminals, data-base systems and data-communication facilities to cope with the need for selective dissemination of information without requiring permanent copies. In today's office environment, however, paper remains a very important carrier of information.

While information access to a computer data base is fast, the flow of paper through regular mail channels is slow. In today's highly competitive business-administration environment, time delays affecting the decision-making process are getting increasingly critical, and one of the key functions envisaged in the conception of tomorrows office is the electronic retrieval of documents in non-coded or image form. Information stored on paper will be scanned, converted into digital form for recording and transmission and presented to the recipient either via display or as hardcopy. The performance of the system when handling information in image form depends largely on transmission channel bandwidth available. While the document transmission delay is in the order of one minute when using CCITT Group III standard digital facsimile transceiver equipment operating over 4800 bit/s lines, it can be made almost instantaneous in a wide-band transport environment.

It seems very likely that the user working in the automated business-administration environment of the future will be able to acquire, process, store, switch, transmit and reproduce documents in Coded Information (CI) or data/text form as well as in Non-Coded Information (NCI) or image form. The combination of LSI technology-based digital computers with modern APA I/O devices will result in integrated data/text and image-processing systems which, together with appropriate signal-processing algorithms, will provide speed, quality, flexibility, convenience and increased productivity. As shown in Figure 1, the functional resources available can be sub-divided into CI, NCI

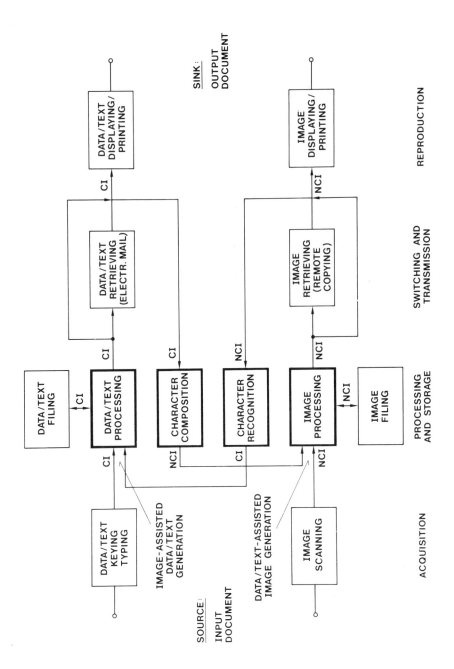

Figure 1. Functional resources for the implementation of office-automation tasks.

and cross-connecting CI/NCI (decoding) and NCI/CI (coding) parts which allow implementation of the following office-automation tasks.

Data/Text Processing

This CI/CI conversion task, typically would assist the user in the creation, correction/revision and presentation of strings of alphanumeric characters. It allows the use of text building blocks to rationalize the production of standard letters with variable insertion of names, addresses, account statements, etc., and can provide support for automated hyphenation, data/text layout, highlighting, tabulating, margin justification and spelling verification.

Functional resources required (* indicates optional):

Data/text keying/typing, processing, filing*, retrieving*, displaying/printing (CI).

Character Composition

In this CI/NCI decoding task, strings of alphanumeric characters are transformed into an array of picture elements for multi-font reproduction with high-resolution APA displays and printers.

Functional resources required (* indicates optional):

Data/text keying/typing, processing, filing*, retrieving* (CI),
Character composition (CI/NCI),
Image processing, filing*, retrieving*, displaying/printing (NCI).

Character Recognition

In this NCI/CI coding task, typewritten or hand-marked documents are scanned and the image data channelled to the character-recognition processor which, after having performed all the necessary formatting, feature detection and classification steps, outputs alphanumeric character code for data/text processing.

Functional resources required (* indicates optional):

Image scanning, processing, filing*, retrieving* (NCI),
Character recognition (NCI/CI),
Data/text processing, filing*, retrieving*, displaying/printing (CI).

Image-Assisted Data/Text Generation

This mixed NCI/CI, CI task provides for human-assisted data/text entry of character-recognition rejects. It is a known fact that the character recognition-performance depends on good print quality of the typed or hand-marked documents. If this requirement is not fulfilled, the character-recognition processor rejects recognition or what is worse, performs a so-called substitution or false recognition. Reliable character-recognition operation is achieved by inhibiting any substitution at the expense of an increased reject rate. To overcome this drawback, APA display devices are used to present the array of picture elements proper to the rejected alphanumermic symbol for subsequent human-assisted data/text entry [3].

Functional resources required (* indicates optional):

Image scanning, processing, filing*, retrieving*, displaying/printing* (NCI),
Character recognition (NCI/CI) and data/text keying/typing (CI),
Data/text processing, filing*, retrieving*, displaying/printing (CI).

Data/Text-Assisted Image Generation

This mixed CI/NCI, NCI task allows merging of character composition and scanned image data for electronic document editing. Capabilities include cropping, size reduction and enlarging, combining of pictures, insertion of captions and comments, etc.

Functional resources required (* indicates optional):

Data/text keying/typing, processing, filing*, retrieving* (CI),
Character composition (CI/NCI) and image scanning (NCI),
Image processing, filing*, retrieving*, displaying/printing (NCI).

Image Processing

This NCI/NCI conversion task, typically would assist the user in the creation, correction and presentation of documents in image form. It allows the user to compand or contrast-enhance, compress/decompress, geometrically transform and process black-and-white and continuous-tone documents for bi-level rendition with APA displays and printers.

Functional resources required (* indicates optional):

Image scanning, processing, filing*, retrieving*, displaying/printing (NCI).

From the above enumeration it becomes obvious that, with the exception of classical data/text processing, all other office-automation tasks require NCI functional resources. In particular, all of them require, in one way or the other, the functional resource of digital image processing, and the purpose of this paper is to discuss the issues of two-dimensional sample-rate conversion and digital screening of continuous-tone image data for bi-level rendition.

TWO-DIMENSIONAL SAMPLE-RATE CONVERSION

General

One of the key motivations in the conception of integrated systems for data/text and image-processing application is the idea of hardware resource sharing. For example, a high-resolution APA printer can be used to reproduce composed text as well as low-resolution facsimile data. However, in order to reproduce the original facsimile in the correct size, its sampling rate must be increased by digital interpolation. Thus, the need for two-dimensional sample-rate conversion becomes apparent. Similarly, if an integrated system is used for editing mixed resolution/size documents, it is necessary to alter the sampling rate for electronic page layout.

The problem of sample-rate conversion in one dimension was examined previously. Schafer and Rabiner [4] treat the problem of interpolation of band-limited signals in terms of the sampling theorem and compare polynomial interpolation with optimal linear filters. It is shown that sampling-rate increase and sampling-rate reduction are basically interpolation processes which can be efficiently implemented using Finite Impulse Response (FIR) digital filters. Furthermore, it is found that sampling-rate conversion between any rational ratio of sampling frequencies can be efficiently implemented by a two-stage process consisting of an integer sampling-rate increase followed by an integer sampling-rate decrease. Oetken, Parks, and Schuessler [5] describe a design of an optimum interpolation consisting of a set of different filters which all together form a symmetric, i.e., linear-phase impulse response at the new sampling rate. Finally, Crochiere and Rabiner [6] suggest a multi-stage design for realizing decimators and interpolators for sampling-rate reduction and sampling-rate increase. They have shown that the process of decimation and interpolation are duals and, therefore, the same set of design considerations applies to both problems.

This section of the paper will discuss the application and extension of the results of earlier findings on sample-rate conversion to the problem of document-size reduction/enlargement. In particular, it will describe in a concise form the possible trade-offs between accuracy, implementation and evaluation of two-dimensional generating interpolation functions for practical scale-change applications.

Ideal Interpolation Filter

Digital document-size reduction/enlargement can be achieved by means of interpolation algorithms which generate image data for sample locations other than the original ones. The filtering action of the sample-rate conversion process is computable by convolving the original image data with an appropriate generating interpolation function.

The general form of one-dimensional interpolation of a band-limited data stream is

$$A'(x) = \sum_n A(n)\, f(x-n),$$

where

$A(n)$: intensity value at original sample location,
$f(x-n)$: generating interpolation function,
$A'(x)$: intensity value at new sample location.

The impulse response of an ideal interpolation filter $f(x) = \sin(x)/x = \mathrm{sinc}(x)$ is regarded as the ideal generating interpolation function. Figure 2 depicts ideal interpolation as a convolution process.

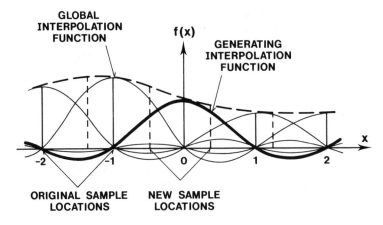

Figure 2. Ideal interpolation as a convolution process.

In order to compute a single element of A', theoretically all
elements of A should be used. In image-processing applications,
however, A is generally a large number and the exact computation of
 A' has to be rejected in order to avoid problems of numerical
stability and cost. In practice, it is advisable to compose the global
interpolation function from generating interpolation functions which are
defined by a small number of elements in a sub-region of the total
interpolation interval.

Design of Interpolation Filters for Practical Applications

First approximation: **Zero-Order or Nearest-Neighbor Interpolation**

The impulse response of an ideal interpolation filter $f(x) = \text{sinc}(x)$
can be approximated by a generating interpolation function $\hat{f}(x)$ of
the form

$$f(x) \approx \hat{f}(x),$$

where

$$f(x) = \begin{cases} 1 & 0 \leq |x| \leq 0.5 \\ 0 & \text{otherwise.} \end{cases}$$

Figure 3 depicts zero-order or nearest-neighbor interpolation as a convolution process. In the two-dimensional case the value at the desired new sample location (k,l), A'$_{(k,l)}$ is determined as the value of (k,l)'s closest original sample location (Fig. 4).

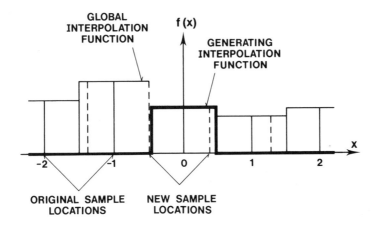

Figure 3. Zero-order or nearest-neighbor interpolation as convolution process.

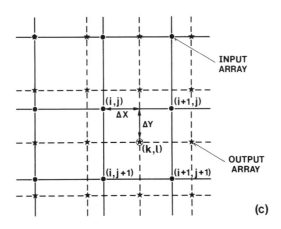

Figure 4. Input and output sampling grids for zero-order or nearest-neighbor interpolation.

$$A'_{(k,l)} = \begin{cases} A_{i,j} & \Delta x < 0.5 \ \& \ \Delta y < 0.5 \\ A_{i+1,j} & \Delta x \geq 0.5 \ \& \ \Delta y < 0.5 \\ A_{i,j+1} & \Delta x < 0.5 \ \& \ \Delta y \geq 0.5 \\ A_{i+1,j+1} & \Delta x \geq 0.5 \ \& \ \Delta y \geq 0.5 \ . \end{cases}$$

Since there are only original sample values assigned to $A'_{(k,l)}$, no arithmetic operations are needed and thus zero-order or nearest-neighbor interpolation is very fast. The original contrast is preserved at the expense of occasional offsets of $A'_{(k,l)}$.

Second approximation: **First-Order or Linear Interpolation**

An improved approximation of the impulse response of an ideal interpolation filter $f(x) = \text{sinc}(x)$ can be obtained by a generating interpolation function $\hat{f}(x)$ of the form

$$f(x) \approx \hat{f}(x),$$

where

$$\hat{f}(x) = \begin{cases} 1 - x, & 0 \leq |x| \leq 1 \\ 0 & \text{otherwise.} \end{cases}$$

Figure 5 depicts first-order or linear interpolation as convolution process. In the two-dimensional case, the value at the desired new sample location (k,l), $A'_{(k,l)}$ is computed as weighted mean of the four original neighbor values. The weights depend on the position of (k,l) relative to the original sample locations (i,j), $(i+1,j)$, $(i,j+1)$, and $(i+1,j+1)$ (Fig.6).

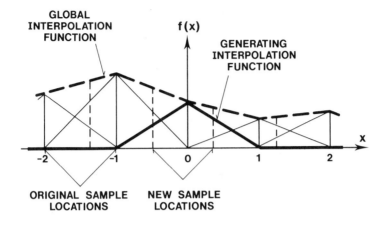

Figure 5. First-order or linear interpolation as convolution process.

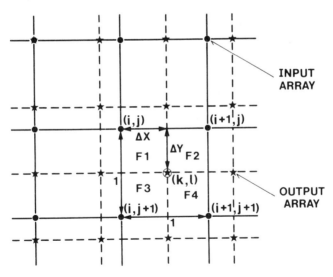

Figure 6. Input and output sampling grids for first-order or linear interpolation.

$$A'_{(k,l)} = (1-\Delta x)(1-\Delta y)A_{ij}$$

$$+ \quad (\Delta x)(1-\Delta y)A_{i+1,j}$$

$$+ \quad (1-\Delta x)(\Delta y)A_{i,j+1}$$

$$+ \quad (\Delta x)(\Delta y)A_{i+1,j+1}.$$

This expression can be modified to reduce the number of multiply-adds from eight to four.

$$A'_{(k,l)} = A_{i,j}$$

$$+ \quad (\Delta x)(A_{i+1,j} - A_{i,j})$$

$$+ \quad (\Delta y)(A_{i,j+1} - A_{i,j})$$

$$+ \quad (\Delta x)(\Delta y)(A_{i,j} - A_{i+1,j} + A_{i,j+1} - A_{i+j,j+1}).$$

First-order or linear interpolation requires the computation of arithmetic functions. There are no offsets of $A'_{(k,l)}$ but a certain image degradation is caused by smoothing effects.

Third approximation: **Third-Order or Cubic-Spline Interpolation**

An even better approximation of the impulse response of an ideal interpolation filter $f(x) = \text{sinc}(x)$ can be obtained by a generating interpolation function $\hat{f}(x)$ of the form

$$f(x) \approx \hat{f}(x),$$

where

$$
\hat{f}(x) = \begin{cases}
|x|^3 - 2|x|^2 + 1 & |x| < 1 \\
-|x|^3 + 5|x|^2 - 8|x| + 4 & 1 \le |x| < 2 \\
0 & \text{otherwise.}
\end{cases}
$$

Figure 7 depicts third-order or cubic-spline interpolation as convolution process. The approximation of $f(x) = \text{sinc}(x)$ by cubic splines was first suggested by Rifman and McKinnon [7]. The cubic-spline-based generating interpolation function extends over four original sample locations. In the two-dimensional case, the value at the desired new sample location (k,l), $A'_{(k,l)}$ is determined as a function of the sixteen original neighbor values (Fig. 8).

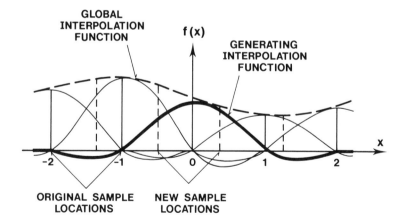

Figure 7. Third-order or cubic spline interpolation as convolution process.

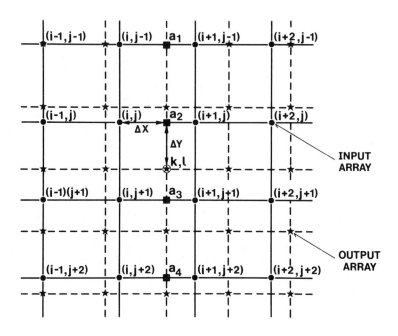

Figure 8. Input and output sampling grids for third-order or
cubic-spline interpolation.

Computation of the cubic convolution in two dimensions is executed as
the product of two one-dimensional cubic convolutions. First, the intensity values at

$$
\begin{aligned}
a_n \quad = \quad & A_{(i-1,j+n-2)} \, \hat{f}(\Delta x + 1) \\
+ \quad & A_{(i,j+n-2)} \, \hat{f}(\Delta x) \\
+ \quad & A_{(i+1,j+n-2)} \, \hat{f}(\Delta x - 1) \\
+ \quad & A_{(i+2,j+n-2)} \, \hat{f}(\Delta x - 2), \qquad n = 1, 2, 3, 4
\end{aligned}
$$

are computed. Second, the sample value at (k,l) A' $_{(k,l)}$ is computed

as cubic convolution of the four column values a_1, a_2, a_3, and a_4:

$$A' _{(k,l)} = a_1 \hat{f}(\Delta y + 1)$$
$$+ a_2 \hat{f}(\Delta y)$$
$$+ a_3 \hat{f}(\Delta y - 1)$$
$$+ a_4 \hat{f}(\Delta y - 2).$$

The straightforward computation of A' $_{(k,l)}$ requires a total of 110
multiply-add operations. In cubic convolution, the output value is determined as a weighted sum of four input values. The weighting factors
are a function of the position of the interpolated value relative to the
input value. By pre-computation and storage of the weighting factors,
the number of multiply-add operations necessary to determine
A' $_{(k,l)}$ is reduced to eight.

Third-order or cubic-spline interpolation requires approximately twice as
much computing compared to first-order or linear interpolation. However, using cubic splines, amplitude and slope continuity properties of
the function to be resampled are preserved and as a consequence the
amount of image degradation is substantially reduced.

The three interpolators discussed so far all satisfy the symmetry
condition $f(x) = f(-x)$ and thus they can be classified as FIR digital
filters with linear-phase characteristics. In this case, group delays are
frequency independent and as such can be compensated to zero phase.
This implies that linear-phase interpolation filters represent the proper
choice for sample-rate conversion of image data, since the human sense
of vision does not tolerate much phase distortion. Figure 9 shows the
filtering action of two-dimensional interpolators. From this spatial frequency representation it can be seen that by increasing the order of

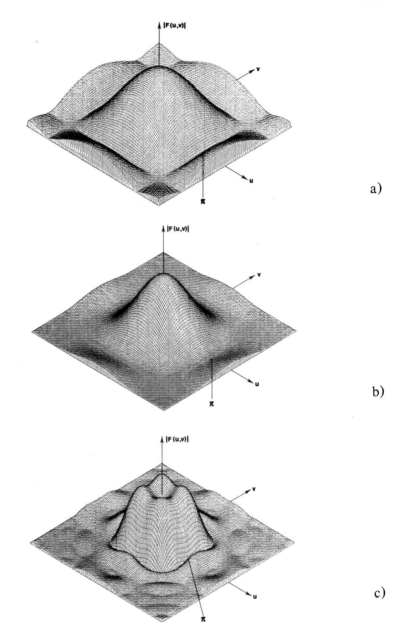

Figure 9. Filtering action of two-dimensional interpolator.
Spatial frequency response | F(u,v) | for:
a) Zero-order or nearest-neighbor interpolation,
b) First-order or linear interpolation,
c) Third-order or cubic-spline interpolation.

the generating interpolation function, a better approximation to the
ideal interpolation filter can be achieved.

Fourth approximation: **Third-Order Polynomial Interpolation (Lagrange)**

In higher-order polynomial interpolation a polynomial of degree $Q - 1$
that passes through Q original sample values $f(x_n)$ is first determined. The interpolated new values $f(x)$ are then computed as samples of this polynomial. Using the Lagrange interpolation formula, $f(x)$
of the polynomial can be obtained direct from $f(x_n)$ [8].

Schafer and Rabiner [4] have interpreted the Lagrange interpolation
formula in terms of $Q - 1$ different impulse responses - corresponding
to the $Q - 1$ different interpolation intervals between the Q original
sample values. They have shown that whenever Q is odd, none of the
impulse responses can have linear phase. However, if Q is even, the
one of the $Q - 1$ impulse responses corresponding to the interpolation
in the central interval does have linear phase. Thus in the Q even
case and interpolation in the central interval, Lagrange interpolation has
no phase distortion and therefore can be used for sample-rate
conversion for image data.

In third-order polynomial interpolation the function extends over
$Q = 4$ equally spaced original sample values, and the Lagrange
formula to determine the interpolated value $f(x)$ is of the form

$$f(x) \;=\; \sum_{\substack{r=1 \\ r \neq s}}^{4} \left(\prod_{s=1}^{4} ((x-x_s)/(r-s)) \, f(x_r) \right),$$

where r and s can take the values 1, 2, 3 or 4.

$$f(x) \quad = \quad f(x_1)\ (x-x_2)\ (x-x_3)\ (x-x_4)/(-6)$$

$$+\ \ f(x_2)\ (x-x_1)\ (x-x_3)\ (x-x_4)/2$$

$$+\ \ f(x_3)\ (x-x_1)\ (x-x_2)\ (x-x_4)/(-2)$$

$$+\ \ f(x_4)\ (x-x_1)\ (x-x_2)\ (x-x_3)/6\ .$$

The two-dimensional Lagrange formula to determine the interpolated value $f(x,y)$ in a grid of $Q_x = 4$ times $Q_y = 4$ equally spaced original sample values is of the form

$$f(x,y) \quad = \quad \underbrace{\sum_{\substack{r=1}}^{4} \sum_{\substack{t=1}}^{4} \left(\prod_{\substack{s=1 \\ r \neq s}}^{4} (x-x_s)/(r-s) \right)}_{f_1(r)} \ \underbrace{\left(\prod_{\substack{u=1 \\ t \neq u}}^{4} (y-y_u)/(t-u) \right)}_{f_2(t)} \ f(x_r,y_t)\ ,$$

where r, s, t and u can take the values 1, 2, 3 or 4.

$f_1(r)$ and $f_2(t)$ are independent for each r and t. They can be treated separately as

$$f_1(1) \quad = \quad (x-x_2)(x-x_3)(x-x_4)/(-6)$$

$$f_1(2) \quad = \quad (x-x_1)(x-x_3)(x-x_4)/2$$

$$f_1(3) \quad = \quad (x-x_1)(x-x_2)(x-x_4)/(-2)$$

$$f_1(4) \quad = \quad (x-x_1)(x-x_2)(x-x_3)/6$$

$$f_2(1) \quad = \quad (y-y_2)(y-y_3)(y-y_4)/(-6)$$

$$f_2(2) \quad = \quad (y-y_1)(y-y_3)(y-y_4)/2$$

$$f_2(3) \quad = \quad (y-y_1)(y-y_2)(y-y_4)/(-2)$$

$$f_2(4) \quad = \quad (y-y_1)(y-y_2)(y-y_3)/6 \ .$$

In order to obtain $f(x,y)$, the sum of all possible products of $f_1(r)$ $f_2(t)$ has to be computed.

In terms of the input and output sampling grids shown in Figure 10, the value at the desired new sample location (k,l), $A'_{(k,l)}$ is determined as a function of the sixteen original neighbor values. Computation of the Lagrange interpolation in two dimensions is executed as the product of two one-dimensional Lagrange interpolations. First, the sample values at

$$
\begin{aligned}
a_n \quad = \quad & A_{(i-1,j+n-2)} \ (\Delta x-1)(\Delta x-2)(\Delta x)/(-6) \\
+ \quad & A_{(i,j+n-2)} \ (\Delta x+1)(\Delta x-1)(\Delta x-2)/2 \\
+ \quad & A_{(i+1,j+n-2)} \ (\Delta x+1)(\Delta x-2)(\Delta x)/(-2) \\
+ \quad & A_{(i+2,j+n-2)} \ (\Delta x+1)(\Delta x-1)(\Delta x)/6 \qquad n = 1, 2, 3, 4
\end{aligned}
$$

are computed. Second, the sample value at (k,l), $A'_{(k,l)}$ is computed as Lagrange interpolation of the four column values a_1, a_2, a_3, and a_4:

$$
\begin{aligned}
A'_{k,l} \quad = \quad & a_1 \ (\Delta y-1)(\Delta y-2)(\Delta y)/(-6) \\
+ \quad & a_2 \ (\Delta y+1)(\Delta y-1)(\Delta y-2)/2 \\
+ \quad & a_3 \ (\Delta y+1)(\Delta y-2)(\Delta y)/(-2) \\
+ \quad & a_4 \ (\Delta y+1)(\Delta y-1)(\Delta y)/6 \ .
\end{aligned}
$$

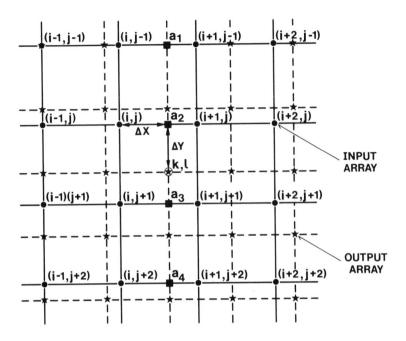

Figure 10. Input and output sampling grids for third-order
polynomial interpolation (Lagrange).

The straightforward computation of $A'_{(k,l)}$ requires a total of 80
multiply-add operations. In third-order polynomial interpolation, the output sample value is determined as a weighted sum of input sample
values. The weighting factors are a function of the position of the interpolated value relative to the input sample values. By pre-computation
and storage of the weighting factors, the number of multiply-add
operations necessary to determine $A'_{(k,l)}$ is reduced to eight.

On Two-Dimensional Sampling

An image $g(x,y)$ sampled with an impulse sampling function $\vartheta(x,y)$
results in a sampled image of the form

$$f(x,y) = COMB_{x,y}[g(x,y)].$$

The spectrum of the sampled image

$$F(u,v) \ = \ (1/|X|) \ (1/|Y|) \ REP_{1/X,1/Y} \ [G(u,v)]$$

can be represented as an infinite repetition of image spectra $G(u,v)$ spaced $1/X$, $1/Y$ cycles apart and weighted by $(1/|X|) \ (1/|Y|)$. Figure 11 shows the topology of $G(u,v)$ repetitions in the spatial frequency domain $F(u,v)$ for the three cases 'oversampled', 'critically sampled' and 'undersampled'. In the last case, image-spectrum folding or aliasing occurs. The effect of aliasing is a distorted image structure which can be expressed in terms of spectral energy. In order to avoid aliasing, the image spectrum $G(u,v)$ has to be low-pass filtered prior to the scale-change operation. The band limitation has to be such as to match the topology of the 'critically sampled' case.

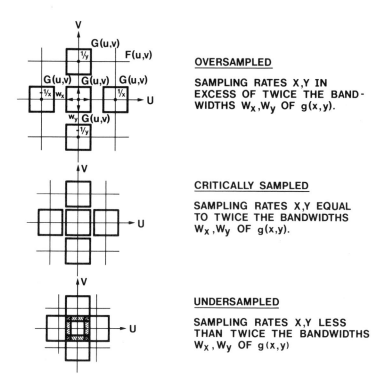

OVERSAMPLED

SAMPLING RATES X,Y IN EXCESS OF TWICE THE BAND-WIDTHS W_x, W_y OF $g(x,y)$.

CRITICALLY SAMPLED

SAMPLING RATES X,Y EQUAL TO TWICE THE BANDWIDTHS W_x, W_y OF $g(x,y)$.

UNDERSAMPLED

SAMPLING RATES X,Y LESS THAN TWICE THE BANDWIDTHS W_x, W_y OF $g(x,y)$

Figure 11. Topology of image spectra $G(u,v)$ in the spatial frequency domain $F(u,v)$.

In practice, a low-pass filter having raised-cosine characteristic is often used. A raised-cosine characteristic consists of a flat amplitude portion and a roll-off portion which has sinusoidal form. The relationship between flat and roll-off portions is specified in terms of a parameter α which is the amount of bandwidth used in excess of the minimum bandwidth w_x, w_y, divided by w_x, w_y [9]. Thus, $\alpha = 1$ is referred to as 100 percent roll-off and means that the total bandwidth used is w_{2x}, w_{2y}, while $\alpha = 0$ indicates that the minimum bandwidth w_x, w_y is employed.

Quantitative Performance Evaluation and Conclusions

The performance of the various two-dimensional digital scale-change algorithms has been quantitatively evaluated for two classes of test signals: scanned continuous-tone text and portrait pictures. In a previous paper [10], the result of a performance evaluation was reported where the test pictures were subjected to a sampling-rate reduction followed by an inverse sampling-rate increase for direct comparison with the band-limited reference images. This approach is considered of limited practical value since it is not very likely that the processes of decimation and interpolation will succeed each other with inverse scale-change ratio. Instead, it is proposed to achieve sampling-rate reduction by simply throwing away band-limited intermediate original sample values prior to any sampling-rate increase. Thus, only the scale-change operation of interpolation will be evaluated. In this case, the procedure for quantitative performance evaluation consists of the following steps:

Select a test picture $f(x,y)$;

Select a scale-change factor defined as $qx = qy = 1/i$, where i is an integer greater than 1;

Compute a band-limited version $f'(x,y)$ of $f(x,y)$ using a raised-cosine characteristic low-pass filter with an α percent roll-off value;

Generate an 'undersampled' representation $s(x,y)$ of $f'(x,y)$ by extracting every i-th pel;

Use any of the above described sample-rate conversion algorithms to reconstruct all i - 1 intermediate pel's omitted in the previous step to produce an approximation $s'(x,y)$ of $f'(x,y)$;

Compare s'(x,y) and f'(x,y) pel by pel to determine, in a mean-square error sense, how well the individual scale-change or resampling algorithms are performing.

A typical set of performance values expressed in [db] is shown in Table I. The results correspond to a two-to-one sample-rate increase using the different interpolation algorithms discussed above and applied to $\alpha = 0.25$ raised-cosine characteristic, low-pass filter band-limited test pictures.

	Continuous-tone text (average) [db]	Continuous-tone portrait (average) [db]
Zero-order or nearest-neighbor assignment	19.5	23.1
First-order or linear interpolation	23.4	33.2
Third-order or cubic-spline interpolation	28.5	33.0
Third-order polynomial interpolation	28.2	34.1

Table I.   Performance in [db] for different sample-rate conversion algorithms (qx = qy = 0.5, α = 0.25).

The performance values expressed in Table I should be inter-preted as a description of the average performance of the interpolator. They are valid for an ensemble of continuous-tone text and portrait pic-tures at the given parameter settings. Naturally, the performance of the sample-rate conversion algorithms changes for new parameter settings and in particular the variation of α does influence the performance substantially. For example, the performance values increase on the average by about 5 db for an ensemble of continuous-tone text and portrait pictures band-limited with a $\alpha = 1$ raised-cosine characteristic low-pass filter. However, for such a setting of α, the smoothing action of the anti-aliasing filter is too strongly degrading image quality.

In general, third-order, i.e., cubic spline and polynomial inter-polation work best for both continuous-tone text and portrait pictures. In particular, they perform substantially better for continuous-tone text than lower-order interpolators. First-order, i.e., linear interpolation performs well for continuous-tone portraits and at lower cost than third-order interpolators. Although zero-order or nearest-neighbor interpolation scores worst for both classes of test pictures, its performance is expected to be substantially higher for truly bi-level imagery.

Pictorial examples illustrating the difference in performance between a zero-order nearest-neighbor and a third-order polynomial interpolator are shown in Figures 12 a) and b) where the sampling rate of a 25 times 25 pel continuous-tone array was increased by a factor of 32 in both the x and y directions. An example of sampling-rate conversion with different scale-change factors in x and y is shown in Figure 13 where an original 100 times 100 pel continuous-tone array (a) is magnified to a 320 times 640 (b) and a 640 times 320 pel array (c), respectively.

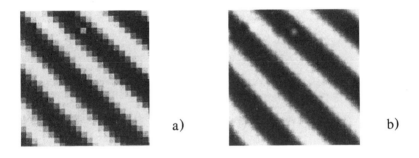

a)   b)

Figure 12. Performance of sampling-rate conversion algorithms.
           a) Zero-order or nearest-neighbor interpolation,
           b) Third-order polynomial interpolation.

Improved performance is achieved at the expense of higher circuit cost and increased processing delay. For many years, fast-logic and high circuit complexity was prohibitive in cost and the implementation of digital image-processing applications had to manage with reduced computational-complexity algorithms which in particular, required as few multiplications as possible. Today however, the situation is different. The constant refining of LSI manufacturing techniques results in further price-performance and package-density improvements. Fast digital multipliers will no longer represent a heavy cost item and it is believed that in the not too distant future, third-order interpolators will be implemented at lesser cost than zero-order interpolators today, thus making it possible to strive for higher image-processing quality.

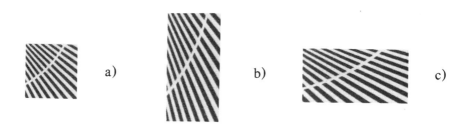

Figure 13. Third-order polynomial interpolation with different
scale-change factors in x and y.
a) 100 × 100 pel continuous-tone array,
b) 320 × 640 pel continuous-tone array,
c) 640 × 320 pel continuous-tone array.

DIGITAL SCREENING OF CONTINUOUS-TONE IMAGE DATA FOR BI-LEVEL RENDITION

General

Integrated systems for data/text and image-processing applications
have to fulfil another basic requirement: The capability to handle continuous-tone achromatic and color pictures. It is a fact that APA computer output printers do not reproduce continuous-tone pictures well.
The reason for this is that the printing process generally exhibits only
two levels of optical density: The presence or absence of ink or toner
particles, for example. A straightforward approach to convert a continuous-tone signal $q(x,y)$ into a bi-level representation is to compare
the signal level at each picture element with a constant-level threshold
$t(x)$ and to make the black or white decision $h(x)$ immediately (Fig.
14). Bi-level representation examples of a constant-level thresholded
continuous-tone portrait picture are shown in Figure 15.

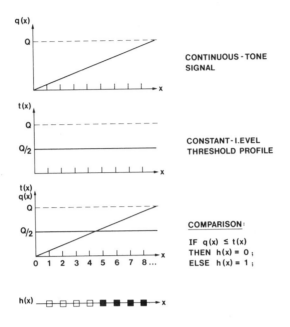

CONTINUOUS - TONE
SIGNAL

CONSTANT - LEVEL
THRESHOLD PROFILE

COMPARISON:

IF q(x) ≤ t(x)
THEN h(x) = 0;
ELSE h(x) = 1;

Figure 14. Principle of constant-level thresholding.

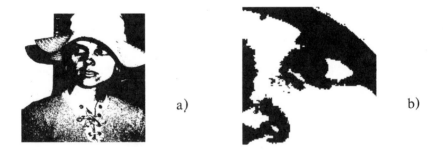

a)

b)

Figure 15. Constant-level thresholding.
a) 600 × 600 pel portrait,
b) 100 × 100 pel cropped and magnified
eye portion of portrait.

Clearly, the tonal rendition is not satisfactory and in order to improve the reproduction fidelity, it is proposed to simulate the process of halftone photography. The latter is used in lithography and photo-engraving and works as follows: When a halftone negative is made from a continuous-tone copy, a halftone screen - in the case of a contact screen an out-of-focus pattern of diffused dots and corresponding spaces - is placed in the light path between the camera lens and a sheet of high-contrast photographic film. The basic function of the half-tone screen and the high-contrast photographic film is to convert the intermediate tones of the original continuous-tone copy into solid dots of equal density but varying size, the centers of which are placed on a regular grid. As the intensity of the reflected light increases from shadows to highlights, the intermediate tones of the continuous-tone copy produce intermediate dot sizes.

In digital halftone photography the function of screening is obtained by increasing the number of threshold crossings of the continuous-tone signal $q(x)$ in a well-defined manner [11, 12]. This is achieved by comparing the signal level at each picture element with a variable-level, repetitive threshold or dither pattern $t[mod(x,i)]$ in order to make the black or white decision $h(x)$. Figure 16 shows the principle of variable-level thresholding in one dimension, and the corresponding examples of bi-level representations are shown in Figure 17.

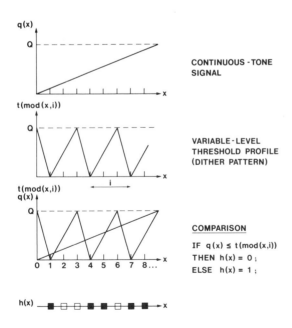

Figure 16. Principle of variable-level thresholding.

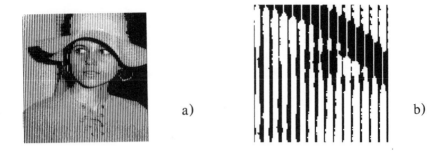

a)

b)

Figure 17. Variable-level thresholding.
a) 600 × 600 pel portrait,
b) 100 × 100 pel cropped and magnified
eye portion of portrait.

In order to break up the width-modulated vertical line structure originating from variable-level thresholding in one dimension, the signal level of each continuous-tone picture element has to be compared with a variable-level threshold profile that varies in the two image dimensions (Fig. 18).

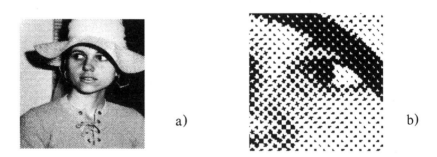

a)

b)

Figure 18. Two-dimensional variable-level thresholding.
a) 600 × 600 pel portrait,
b) 100 × 100 pel cropped and magnified
eye portion of portrait.

A comparison of Figures 15, 17 and 18 shows the improved rendition obtained with two-dimensional variable-level thresholding. The penalty paid for this improvement is a certain loss in spatial frequency and the main issue encountered in digital screening is the trade-off between tonal fidelity and the resolution of pictorial detail.

Digital Screening

The purpose of digital screening is to divide the total image area into a regular pattern of unit areas and to represent different shades of gray by printing a corresponding, selected pattern of fixed-size black dots for each unit. In practice, the continuous-tone signal is first quantized into an appropriate number of amplitude levels $q(x,y)$ and then compared against a repetitive threshold profile that varies in two dimensions $t[\text{mod}(x,i), \text{mod}(y,j)]$ to generate a bi-level image value $h(x,y)$ such that for $q(x,y) \leq t[\text{mod}(x,i), \text{mod}(y,j)]$ $h(x,y) =$ '0' or white and '1' or black otherwise (Fig. 19).

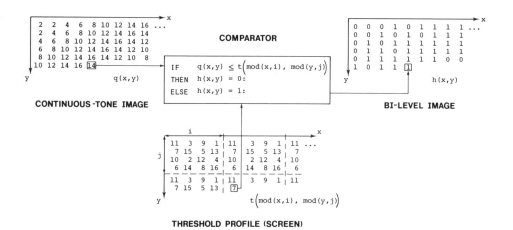

Figure 19. Principle of digital screening.

Over the past years, several design approaches to optimize the threshold profile with respect to spatial resolution and tonal fidelity have been proposed [13, 14, 15]. Typically, an optimized threshold profile consists of a set of threshold values as shown in Figure 20.

22	6	18	2	21	5	17	1
14	30	10	26	13	29	9	25
20	4	24	8	19	3	23	7
12	28	16	32	11	27	75	31
21	5	17	1	22	6	18	2
13	29	9	25	14	30	10	26
19	3	23	7	20	4	24	8
11	27	15	31	12	28	16	32

Figure 20. Optimized threshold profile.
$(1 \le q(x,y) \le 32)$.

This arrangement of threshold values can be characterized as one in which the i-th and the (i+1)-th threshold values are placed as far as possible from one another in the threshold profile. Digital screens fulfilling this characteristic generate dispersed dot patterns having the capability of reproducing high spatial resolutions at an ideally linear density versus amplitude-level relationship. For example, the tonal mid-range reproduced under ideal conditions corresponds to a checkerboard-like arrangement of squared dots as shown in Figure 21 a). For comparison, the reproduction of the tonal mid-range with an non-ideal on/off printing device producing overlapping dots, as it may occur in ink-jet printing or electrophotography, is shown in Figure 21 b).

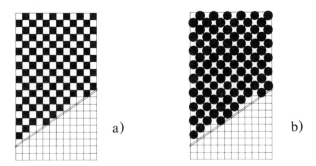

a) b)

Figure 21. Dispersed dot-pattern generation.
a) Ideal rendition,
b) Non-ideal rendition.

Clearly, the density versus amplitude-level relationship becomes dot-pattern dependent, i.e., non-linear and therefore cumbersome to control (Fig. 22).

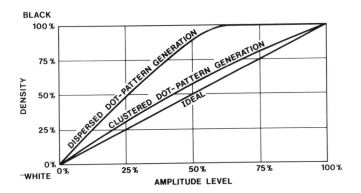

Figure 22. Average plots of density versus amplitude relationships for dispersed and clustered dot-pattern generation and non-ideal rendition (i.e., circular-shaped dots of $\pi/2$ unit area).

Very often though, a linear density versus amplitude-level relationship is required in hardcopy reproduction with non-ideal printing devices, and in practice, this requirement is closely fulfilled by clustered dot patterns as shown in Figures 23 a) and b).

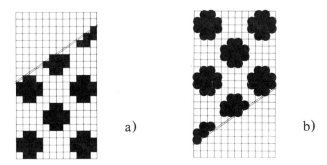

Figure 23. Clustered dot pattern generation.
a) Ideal rendition,
b) Non-ideal rendition.

Dot overlap is naturally compensated at the interior of the clustered dot patterns and a close-to-linear density versus amplitude-level relationship is achieved at the expense of a certain loss in spatial resolution. The threshold profile to generate clustered dot patterns can be characterized as one in which the i-th and the (i+1)-th threshold values are placed as closely as possible to one another. Figure 24 shows an example of a threshold profile to generate clustered dot patterns under 45º, an angle at which the cut-off bandwidth for binocular vision is reduced by 10 - 20 % as compared to horizontally and vertically oriented gratings [16].

19	25	23	17	14	8	10	16
21	31	29	27	12	2	4	6
28	30	32	22	5	3	1	11
18	24	26	20	15	9	7	13
14	8	10	16	19	25	23	17
12	2	4	6	21	31	29	27
5	3	1	11	28	30	32	22
15	9	7	13	18	24	26	20

Figure 24. Typical threshold profile for clustered dot-pattern generation $(1 \leq q(x,y) \leq 32)$.

Analytic Procedure to Design Clustered Dot Patterns

A straightforward approach to design clustered dot patterns is to determine the clustering of thresholded values according to the function $x^2 + y^2 = r^2$, where x, y are Cartesian coordinates and r the radius of a circle. Figure 25 a) shows a gray-scale field demonstrating

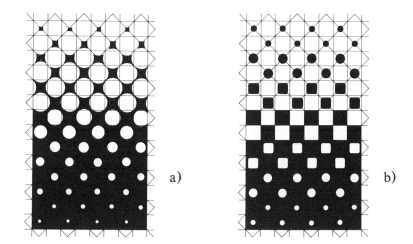

Figure 25. Examples of gray-scale fields.
      a) Different dominant pattern shapes in the dark and
         highlight areas,
      b) Same dominant pattern shapes in the dark and
         highlight areas.

the effect of increasing the radius r from small values (bottom range)
to medium values (mid-range) to large values (top range). In the
bottom and mid-range parts of the gray scale, circular-shaped white patterns are dominant. The radius r eventually exceeds the value where
it breaks up the black areas and gradually, hyperbolic square-shaped
black patterns become dominant. Experiments have shown that the
visibility and the perception of different shape dominant patterns may
be subjectively disturbing in bi-level rendition of continuous-tone data.

Figure 25 b) shows a gray-scale field in which the dominant
white and black patterns have the same shape. The dominant patterns
in the dark area of the gray-scale field are circular-shaped white patterns which towards the intermediate tones gradually change their
shape to a square. Similarly, the dominant pattern in the white area of
the gray-scale field are circular-shaped black patterns which towards the
intermediate tones gradually change their shape to a square. The point
then is to find an analytic procedure to determine the clustering of
threshold values such that the shape of the dominant pattern smoothly
transitions from a circle to a square and back to a circle with increasing amplitude levels. The approach taken is based on the concept
of 'Super-Circles' or geometric figures of the general form
$x^{2n} + y^{2n} = r^{2n}$, where $1 \leq n \leq \infty$. There are two parameters:
 n determines the shape of the figure and together with the radius r
the area A (Fig. 26).

$$A = 4r^2 [\Gamma(1/2n + 1)]^2 / \Gamma(1/n + 1), \qquad \text{where}$$

$\Gamma:$ Gamma function.

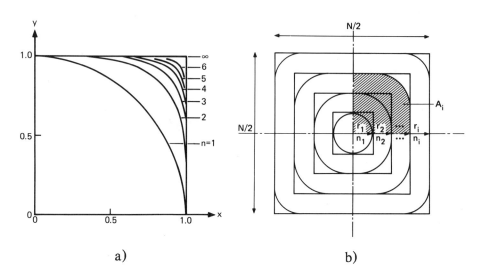

a)   b)

Figure 26. 'Super-Circle' dot-pattern generation.
a) Examples of first quadrant plots
$1 \le n \le \infty$, r = constant.
b) Area A_i as a function of r_i and n_i.

There are two limiting cases of special interest:

$n = 1$: the conventional circle with area πr^2,
$n = \infty$: the square with area $4r^2$.

The procedure to determine a set of n and r values for the computation of 2Q discrete 'Super-Circle' patterns with areas in multiples of C = constant is as follows: First, the value r as a function of the area A is determined for both circles and squares. It is assumed that for the first Q 'Super-Circles' of the set, a circle design is a satisfactory approximation. For the Q larger 'Super-Circles' the strategy consists in choosing a reasonable set of r values and to fit the area profile by varying n. This is obtained by making a linear transition from the n = 1 curve (circles) to the n = ∞ curve (squares) using the formula

$$r_{chosen}(A) = r_{circle}(A) - ((A - Q)/Q) \, [r_{circle}(A) - r_{square}(A)] \, ,$$

where A is the area in units of C = constant. Finally, the corresponding value for n is computed for given radius $r_{chosen}(A)$, area A and Gamma function Γ.

For illustration purposes, Figure 27 a) shows the resulting set of 16 equal area increment 'Super-Circle' patterns plotted on a 128 × 128 pseudo-analog grid. For use in low-resolution halftone printing, it is desired to spatially quantize the set of 'Super-Circle' patterns. For example, the set of 16 'Super-Circle' designs lends itself for distribution in a 4 × 4 quantization grid with an increment of one print position per level. The basic scheme is to proceed through the set, turning on the print position which covers the greatest area of 'Super-Circle' of those that have yet to be turned on. The implementation of the spatial quantization procedure consists of integrating by counting the positions covered by the 'Super-Circle' pattern in the 32 × 32 subarray corresponding to the 4 × 4 quantization grid. In the case where several subarrays are identical by symmetry, the procedure selects the print position in the 4 × 4 quantization grid in a consistent predetermined sequence. Having exhausted all symmetries, the procedure selects the next of the image subarrays, etc. The spatially quantized set of 4 × 4 'Super-Circle' patterns are shown in Figure 27 b) for ideal rendition, and in Figure 27 c) for non-ideal rendition. Finally, the corresponding 'Super-Circle' threshold profile is shown in Figure 28.

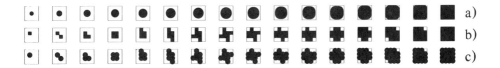

Figure 27. Sixteen equal area increment 'Super-Circle' patterns.
   a) 128 × 128 pseudo-analog plots,
   b) Spatially quantized to 16 print positions
    (ideal rendition, i.e., square imprints),
   c) Spatially quantized to 16 print positions
    (non-ideal rendition, i.e., circular imprints).

19	25	23	17	14	8	10	16
21	31	29	27	12	2	4	6
28	30	32	22	5	3	1	11
18	24	26	20	15	9	7	13
14	8	10	16	19	25	23	17
12	2	4	6	21	31	29	27
5	3	1	11	28	30	32	22
15	9	7	13	18	24	26	20

Figure 28. 'Super-Circle' threshold profile for 32 amplitude levels.

The threshold values contained in quadrants I and III will generate 16 black 'Super-Circle' patterns in transition from small circles to a square and arranged under an angle of 45°, while the threshold values contained in quadrants II and IV will generate 16 white 'Super-circle' patterns in transition from a square to a small circle and arranged under an angle of 135°.

The 'Super-Circle' threshold profile shown in Figure 28 has the capability to render 32 amplitude levels, a dynamic range typically required in computer-output image printing. In order to reproduce 32 amplitude levels with a screening resolution of 70 clusters/inch corresponding to a screening resolution typically encountered in newspaper printing, a printing resolution of approximately 400 imprints or pels/inch is necessary. Magazine-quality printing will require a screening resolution of approximately 100 clusters/inch, bringing the necessary printing resolution close to 600 imprints or pels/inch. Naturally, the analytic properties of the 'Super-Circle' concept will allow the computation of any size threshold profile. In particular, it also allows 'Super-Ellipse' threshold profiles to be conceived, a feature which is of particular interest whenever the printing resolution is different in x and y. Finally, the 'Super-Circle' concept can also be used to compute clusters placed under a certain tilt angle α. In this case, the coordinate pair x_i, y_i of the i-th cluster center may no longer coincide with the quantization grid for printing. As a result, the spatial quantization procedure as described above has to be repeated for each 'Super-Circle' cluster with the result that the threshold values computed no longer repeat themselves periodically (Fig. 29). Threshold profile computation for tilted 'Super-Circle' halftone screens find their application in digital color image reproduction.

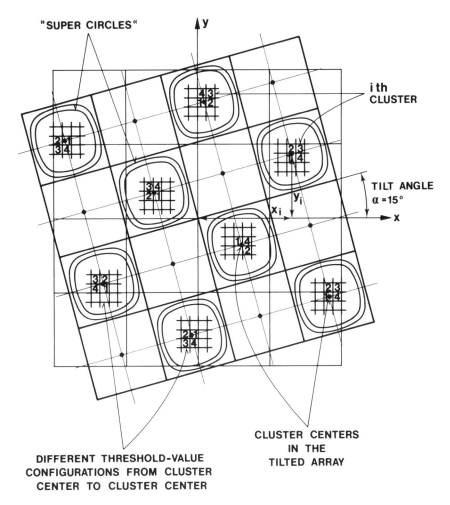

Figure 29. Principle of threshold profile computation for tilted 'Super-Circle' halftone screens.

Application of 'Super-Circle' Halftoning to Digital Color Image Reproduction and Conclusions

To implement a color-reproduction process, the original color picture is first bandpass filtered into its additive primaries green, blue and red. The color extractions are then screened and for their reconstruction on white paper, the subtractive primaries magenta (green absorbant), yellow (blue absorbant) and cyan (red absorbant) are used. The superimposed halftone representation tends to produce low-frequency and subjectively disturbing Moiré patterns when the angular separation between the individual screens is small. Rather than to try to overprint the individual color screens under the same orientation, the angular separation between them is made as large as possible. In four-color reproduction, the subtractive primaries magenta and cyan as well as the synthetically generated black extraction are printed at $\alpha = 15^O$, $\alpha = 75^O$ and $\alpha = 45^O$ allowing a 30° angular separation between them. The subjective primary yellow is printed under an angle of $\alpha = 90^O$. Tilted 'Super-Circle' threshold profiles capable of rendering up to 32 amplitude levels have been computed and used to screen the four color extractions of the portrait picture as shown in Figures 30 and 31. The individual tilt angles α can easily be depicted from the corresponding magnitude plots of the Fourier transform $|F(u,v)|$. Finally, a cropped and magnified portion of the same portrait picture is shown as color reproduction in Figure 32.

The design of 'Super-Circle'-based threshold profiles for halftone screening has been outlined and its application to achromatic and color image data demonstrated. The analytic nature of the 'Super-Circle' concept allows the computation of any size threshold profile and at any tilt angle α. The implementation in fast digital logic anticipates its usage in conjunction with high-resolution, non-ideal computer-output image printing devices.

Acknowledgement

The author would like to thank S. Lomecky, University of Zurich, B. P. Medoff, Stanford University and N. M. Eisman, M.I.T. Cambridge, for their many suggestions and help towards the software implementation of the algorithms discussed in this paper.

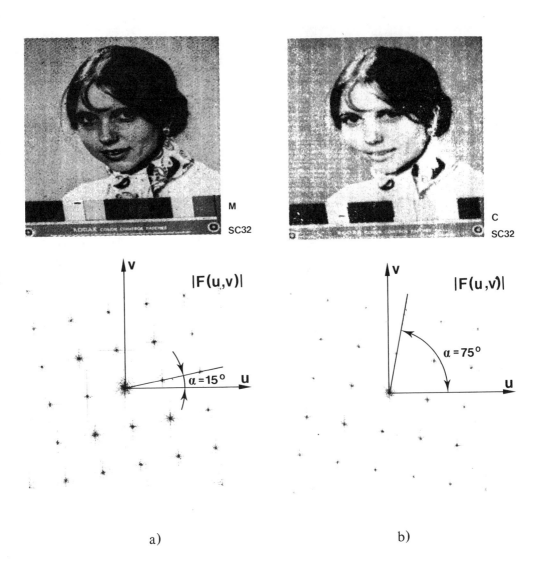

a)   b)

Figure 30. 'Super-Circle' halftoning.
a) magenta screen,
b) cyan screen.

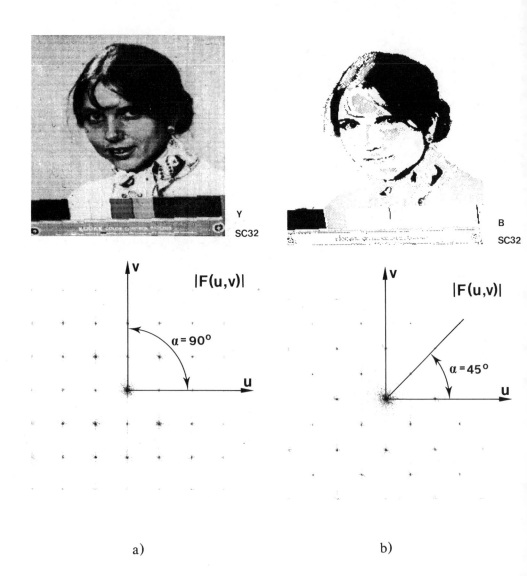

a)   b)

Figure 31. 'Super-Circle' halftoning.
a) yellow screen,
b) black screen.

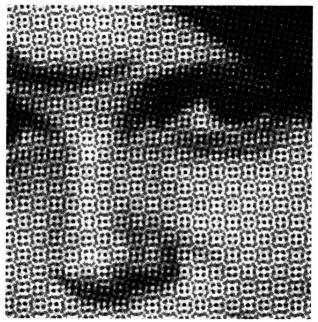

Figure 32. 'Super-Circle' halftoning: Example of digital color image reproduction (cropped and magnified portion of portrait picture).

REFERENCES

1. R. D. Compton, 'The Solid-State Imaging Revolution', Electro-Optical System Design, April, 1974.

2. C. Machover, M. Neighbors and C. Stuart, 'Graphic Displays', IEEE Spectrum, vol. 14, no. 8, 1977.

3. H. Falk, 'Electronics Prints Magazine', IEEE Spectrum, February 1976.

4. R. W. Schafer and L. R. Rabiner, 'A Digital Signal Processing Approach to Interpolation', Proc. IEEE, vol. 61, no. 6, June 1973.

5. G. Oetken, T. W. Parks, and H. W. Schuessler, 'New Results in the Design of Digital Interpolators', IEEE Trans. on Acoustics, Speech and Signal Processing, vol. ASSP-23, no. 3, June 1975.

6. R. E. Crochiere and L. R. Rabiner, 'Optimum FIR Digital Filter Implementation and Narrow-Band Filtering', IEEE Trans. on Acoustics, Speech and Signal Processing, vol. ASSP-23, no. 5, October 1975.

7. S. S. Rifman and D. M. McKinnon, 'Evaluation of Digital Correction Techniques for ERTS Images', TRW Report, No. 20634-6003-TU-00, March 1974.

8. E. Isaacson and H. B. Keller, 'Analysis of Numerical Methods', J. Wiley and Sons, Inc., New York, 1966.

9. R. W. Lucky, J. Salz, E. J. Weldon, Jr., 'Principles of Data Communication', McGraw-Hill, New York, 1968.

10. P. Stucki and S. Lomecky, 'Digital PEL-Size Transformations', Informations- und Systemtheorie in der digitalen Nachrichtentechnik, NTG Fachberichte, Band 65, VDE Verlag, Berlin, 1978.

11. P. Stucki, 'Statistical Measurements for Television Picture Classification and Some Experiments in Video Signal Encoding', DIC Thesis, Imperial College of Science and Technology, London, July 1968.

12. J. O. Limb, 'Design of Dither Waveforms for Quantized Visual Signals', B. S. T. J. 48, 1969.

13. B. E. Bayer, 'An Optimum Method for Two-Level Rendition of Continuous-Tone Pictures', International Conference on Communications, Conference Record, CAT. No. 73 CHO 744-3CSCB, June 1973.

14. P. Stucki, 'Comparison and Optimization of Computer Generated Digital Halftone Pictures', SID International Symposium Digest of Technical Papers, Vol. VI, April 1975.

15. J. F. Jarvis, C. N. Judice and W. H. Ninke, 'A Survey of Techniques for the Display of Continuous-Tone Pictures on Bi-Level Displays', Computer Graphics and Image Processing, 5, 1976.

16. G. C. Higgins and K. Stultz, 'Visual Acuity as Measured with Various Orientations of a Parallel-Line Test Object', J. Opt. Soc. Am., 38, No. 9, 1948.

IMAGE PROCESSING AND COMPUTER GRAPHICS

Robin Williams
IBM Research Laboratory
San Jose, California 95193

ABSTRACT

For a number of years, image processing and computer graphics have been treated as separate subjects. Conferences and journals exist for one group or the other but rarely for both groups. However, the hardware for computer graphics, and even for alphanumeric displays, more and more frequently use raster-scan processing. Likewise, image processing equipment can be used for drawing vectors and writing text as well as outputting images. Recently, therefore, there has been interest in combining interactive-graphics techniques and image-processing techniques on the same display screen. This paper explores the use of 'image mixing' techniques and its uses in graphics and the corresponding use of graphics in image processing. Several examples are explained.

COLOR-GRAPHICS TERMINAL

As a result of some earlier work on decision making using graphic displays of data on maps, a need arose in our laboratory for a terminal that could display vectors and text in color. We wanted to be able to distinguish between two or more similar maps (line drawings) and display numerical values on these maps. Techniques using dotted lines or different gray scales were not good enough, hence the need for color. Also, we wanted to be able to display images derived either from computation or from a video scanner operating on-line in real time.

Few display technologies are suitable for making a color-graphics terminal. By far the simplest, cheapest, and most versatile is the cathode-ray-tube (CRT) display [1]. There is a vector-beam-deflection CRT system that uses beam penetration into multiple layers of of phosphors [2]. This technique provides a few colors (green, yellow, orange, and red). However, only a raster-scan color-TV display is capable of providing a full range of colors. Recently, several raster-scan display systems, consisting of a display processor, a refresh memory and a CRT TV monitor, have appeared on the market [3]. The display processor contains a vector generator and a character generator that generate lines or characters in the form of bit patterns which are then stored in the refresh memory. The refresh memory logically consists of several two-dimensional bit planes. Corresponding bits in each plane control the color/intensity of a corresponding picture element (pel) on the display screen. Thus a system with three planes can be used to provide on/off control of the red, blue and green pels; the display processor cycles through the refresh memory to paint the images in the refresh buffer onto the TV screen.

We assembled a graphics terminal which we called 'RAINBOW' using a raster-scan CRT display, a mini-computer and several input devices (Figure 1). Our first display device had seven refresh-memory planes, 512 x 512 bits each, and a video look-up table sometimes called a 'color mapper' (Figure 2). The display processor was modified to allow the vector/character generators to simultaneously generate bit patterns in any combination of the seven planes of memory. Each pel position has effectively a seven-bit number to store color/intensity information. This number is used to address a table (12 bits wide) that contains the actual color/intensity specifications. Four bits specify the brightness for each of the green, red, and blue signals giving a possible 16 shades for each primary color, or 16 x 16 x 16 = 4096 possible colors. The programmer (or end-user) can specify the contents of the video look-up table (VLT) and hence assign a color for each of the seven-bit pel values. Thus, one can choose any 128 colors from the 4096 possible colors. This provides extremely flexible color specifications.

Currently, the most expensive component in raster-scan display systems is the refresh memory. Seven planes of memory are expensive. We built two more experimental display systems, one with three refresh memory planes and one with four. Three seems to be a reasonable number for most graphics applications and one can still use a VLT to get eight out of two to the power n possible colors for flexibility, where n is the number of bits per word in the VLT. Four planes of 512 x 512 bits each was chosen for one system so that it could be used as a monochromatic display with one plane of 1024 x 1024 bits

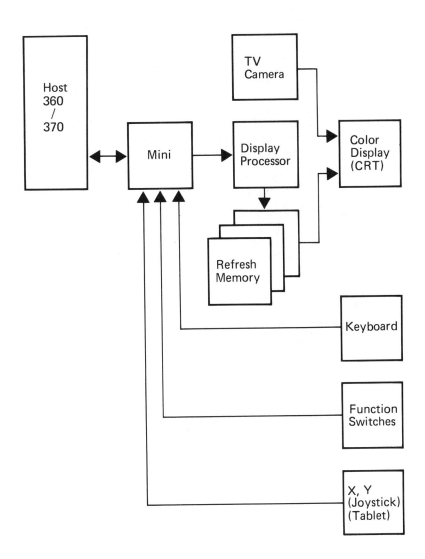

Figure 1. The display hardware.

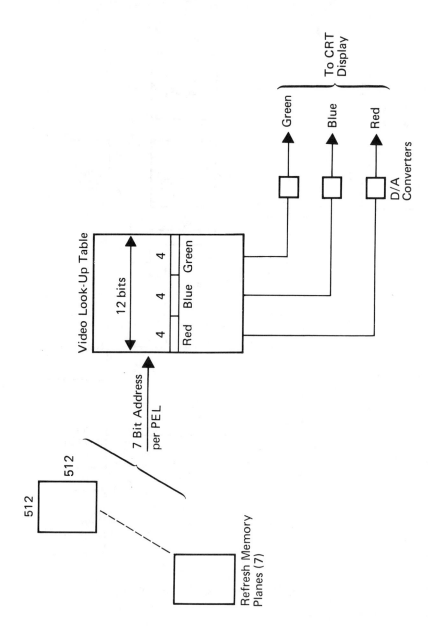

Figure 2. Video look-up table organization.

or as a color display with four four planes of 512 × 512 bits. With more memory, one can have a higher-resolution full-color system. Color CRTs with sufficient shadow-mask resolution are made [4], but there is difficulty in obtaining good convergence at higher resolutions.

Raster-scan display systems are very flexible. For example, we have used a TV camera as a scanner to provide an image mixing capability. The camera signal is fed through a mixer to one of the red, green, or blue inputs of the TV monitor. Users of the graphics terminal can place objects in front of the TV camera and display a scanned image of the object and then draw lines or write text on the same screen under control of the computer. We expect that TV cameras will soon be available with at least 512 × 512 resolution using CCD arrays. These scanners promise to be cheap, reliable, stable and with much less distortion than current TV cameras. The display memory-scanner technology is very suitable for LSI techniques.

Because the display system is compatible with television technology, we can also easily connect several monitors in parallel as slave screens and we use a projection system onto a large screen for demonstrations, group meeting, etc. The complete color, image-mixed signal is displayed in each case.

For hardcopy output, we are constructing an interface between the refresh memory and raster out printer/plotters. The data is already organized in a form that is very close to that required by raster output devices which is another advantage of this type of display system. There are many black/white raster-output devices and one color ink-jet device was demonstrated by JET-AB, Lund, Sweden in 1974 (5).

The first 'RAINBOW' terminal used an IBM System 7 computer as a controller for a modified display processor and for handling the peripheral I/O devices. Although a general-purpose mini-computer was used for convenience in programming, a microcomputer or special-purpose hardware could be used. The two newer display terminals use the IBM 5100 computer and a different display processor. The main advantage of these systems over our first terminal is that they are portable and less expensive. The System 7 has convenient digital and analog I/O interfaces which were used to interface the display processor, keyboard, joystick, local-control switches and data-tablet. The mini is programmed to communicate via an RS232 telecommunications interface with a host (S/360 or S/370 computer, at speeds ranging from 110 to 9600 baud (bits/sec). It performs buffering and formatting of data received from the host, and then outputs commands to the display processor. Inputs from the keyboard, joystick, etc. may cause output of commands to the display processor, communications to the host, and/or control of the cursor generator. Our graphic support software is written so that all our application programs written for storage tube terminals or for IBM 2250 or 3250 displays, can be run on the color terminal without changing the application code.

SIMILARITIES AND DIFFERENCES BETWEEN COMPUTER GRAPHICS AND IMAGE PROCESSING

General Observations

Computer graphics generally involves synthesis techniques. A display terminal is used to input x, y points and draw lines to specify complete drawings. Occasionally, mathematical functions are used to calculate arcs, radii, distances, etc., and topological constraint-handling functions are used to force objects to be connected. 3D objects are input by turning an object on its side (by software or special hardware) and then by drawing the side views. Thus, one is trying to represent objects in a computer form that allows the objects to be viewed and analyzed easily. Examples are to draw a map and calculate stresses, volumes and weights. In some cases, the picture itself is the end result, for example, in making the figures for this paper. Figure 3 shows a mapping example.

Image processing by contrast is mostly analysis and numerous techniques for analyzing an image in order to extract information from it have been devised and studied [6]. An image is somehow scanned and stored in the computer memory. It may be thought of as stored as a matrix of pels (picture elements). Each pel may be from one to n bits and may represent gray scale in black and white or color. Programs then process the image; for example, to find the outline of regions such as lakes in an image of the earth's surface, to determine the features of a character for character recognition, or to analyze the contents of an aerial photograph to determine the condition of wheat in a wheat-growing area. Much work has been expended on algorithms and heuristics for such tasks with the overall goal being automatic image understanding. Pictures can be the end result of image processing too, for example, in image enhancement an image is processed to enhance the contrast in the image. For example, LANDSAT images from satellites are processed to correct for geometric distortions; medical x-ray images are processed to produce enhanced gray-scale images to aid human study. Figure 4 shows an example of image output.

Hardware Systems

For computer-graphics displays, several types of techniques have been used [7]. These include random-scan or beam-deflection CRTs, storage CRTs, raster-scan CRTs. It is the move toward raster-scan technologies, made possible by cheaper memory prices and new technologies (liquid crystal), that make it possible to display images as well as vectors and text on the same screen. Many output printers also use raster-scan techniques [8]. Input scanners by their nature are raster-scan devices (although a few special-purpose scanners have been made that are not simple raster-scan devices).

Image-processing displays are raster-scan devices (generally CRTs). A long time ago, most image output was encoded and output very crudely on line printers, but now refreshed displays and raster printers are commonly used. Film is used for static output or for final form. As with computer graphics, regular RAM memory is often used as a refresh store for CRT raster-scan displays and so the technologies used for graphics and image processing are now very similar. By just adding a vector and text generator to image-processing equipment and an input xy device, one can do fully interactive graphics too. Likewise, in a graphics system using frame buffers, by just adding a sequential data path to the refresh memory, one can display images with reasonable speed on the same equipment. Hence, the obvious mixing of computer graphics and images.

Software and Data Structures

In computer graphics systems, graphical objects are represented in a variety of ways in a database. Frequently, data is represented as a hierarchy of line drawings and is highly structured to represent geometric and topological relationships among objects. To make processing efficient in applications where speed is important, the data is mapped into other specialized structures to represent pictures displayed on a screen. With some kinds of hardware, a third representation of the data is frequently needed in the form of a display file for the hardware to refresh the CRT [9]. In beam-directed CRT systems, the display file is a sequential set of graphic orders for drawing lines and for outputting text, whereas in a raster system the refresh mechanism is from a bit buffer. Some raster systems have used a list of graphic orders and text strings when the hardware was fast enough to generate the bit data during the raster scan. Storage tube displays do not require such a display file, but for compatibility among display systems some people use display files anyway.

Software packages, language extensions and even special-purpose languages have been created for graphics [10]. There are over 50 similar packages (only differing in detail). They allow one to output lines and text, input points and text, and identify entities on a screen by pointing to them. Entire or partial drawings may be scaled, rotated and positioned anywhere on the screen, even in real time (animation) and perspective views of objects can be created. There have recently been efforts to standardize these techniques and proposals exist for standard graphics software.

Image processing, by and large, employs simpler data representation: planes of bits, or arrays of pels. The display of the data on a raster printer or a CRT also is in the same form. The functions required of the software to display and process the data are pel or area oriented and quite different from that required for graphics. Thus, image-processing software packages have also been created, but 'pure graphic software' and 'pure image software' [11] have virtually nothing in common. Only where mixed techniques are employed (drawing lines on images, outputting images with drawings) is one likely to find any overlap.

Generally speaking, interactive graphics applications (except for simple plotting of graphs) requires very sophisticated programming techniques to build, manipulate and edit data structures representing graphical data and interrelationships between objects. Image processing is simpler in this respect but requires sophistications of quite a different nature (namely, analytical recognition heuristics).

COMPUTER GRAPHICS USING IMAGES

Background Overlay

Raster-scan graphics system, whether color or black and white, allow one to display images as well as graphics. Thus, it is natural to see what can be gained by adding images to traditional graphics applications. The most obvious thing to do is to add backgrounds to graphic displays. Thus, when a street map is being displayed, data other than streets (hospital symbols, parks, railways, rivers, etc.) can be added by scanning and displaying a regular paper map of the same area with all this extra information on it, see Fig. 5. on it. If the scanner works in real time, such as a TV camera, then the image from the scanner can be mixed electronically with the signal to the display monitor and the

Figure 5. An image-mixing example.

Figure 3. An example of graphics output.

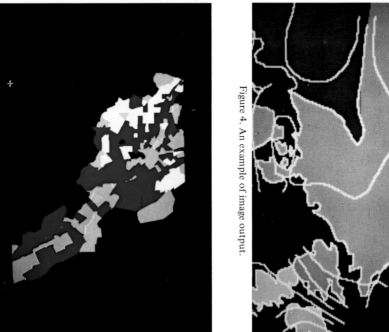

Figure 6. Area filling in computer graphics.

Figure 4. An example of image output.

resultant composite signal (street map and scanned map) can be displayed. The maps can be registered with respect to each other by adjusting the camera position (which is difficult) or by scaling and shifting the computer data and redisplaying it. One method to register the computer data with a scanned image is to identify three points on the image and then three corresponding points on the line drawings, from which the proper position, size and orientation of the graphic output can be calculated. The graphical data is then redrawn and correctly superimposed on the image. When a real-time image cannot be created, it is possible to scan the image, store it in the computer and output it as an image. Of course, this requires extra storage and also slows down program-execution speed considerably.

The background-image idea is quite useful in practice. We have shown that such a technique can be used for map editing and map digitizing with gains of two to ten times in overall productivity compared.

Other applications include the genration of business forms, the input of data on a form, decision making where an image provides helpful background data (e.g., utility mapping, planning for new telephone cables, etc.) and artwork generation. In general, background images are useful for providing reference information, for aiding input of points, lines or text, or for generating composite pictures to photographs or output.

Graphic Primitives

Images can also be used as additional primitives in graphics systems. Some pictures are best represented as bit patterns (e.g., company logos, map symbols and photographs), and in our Picture Building System, for example, such images can be scaled, rotated and positioned just like lines and text [12]. The use of images expands the range of applications, or the range of functions for exisiting applications in obvious ways. A most useful application of this technique is for defining new character sets or for representing different printer-type fonts.

Shading and Area Filling

A third major use of image (or bit-processing techniques) is in the generation of shaded pictures. Starting from line drawings, areas can be filled in with colors, cross-hatching or various other patterns, see Fig. 6. Thus, a 3D picture may be shaded to make it appear more real; one can even add highlights and shadows [13, 14]. This is useful when the end result is a picture (e.g., a new building design or a new car body, or art for publications). Pattern filling is useful for the design of materials, carpets and drapes, wallpapers and clothes.

Realism and Animation

There is also an interest in creating realistic-looking graphic output of near photographic quality. Images are synthesized with shading and illumination properly rendered [13, 14]. Sometimes the output is recorded on film, slightly altered, recorded again, etc., to make an animated movie [15]. A dream of computer-graphics workers was to create films that are so realistic as to be indistinguishable from the real thing, and then furthermore, to produce animation that runs in real time. Such animation is useful for preparing people for difficult exercises before undertaking the real exercise (e.g., training pilots, landing a lunar module and driving a car); for showing otherwise impossible things such as how a downtown area will look if a certain proposed building is built, or for training and assisting doctors and surgeons with animated movies of medical operations. The big problem, of course, is processing vast amounts of information representing images. In graphics, special hardware for line drawings and text was built by several companies to rotate, scale, shift and clip 3D 'wire-frame' lines in real time, and equipment was built to aid the processing of color-shaded images [16].

IMAGE PROCESSING USING COMPUTER GRAPHICS

Feature Extraction

Workers in the image-processing field have tackled the problems of extracting 'features' from images [6]. A feature can range from finding a straight edge, to identifying a 3D solid object to classifying a pattern. Frequently, the feature is then represented in a new way different from the original image, as vectors, as text or in special encodings. It is then possible to show the feature as an overlay on the image, either in a different brightness or color if a display is used. Image processing can become interactive too and human operators can observe the features found in the image and direct the computation to test for other features or cause the features to be adjusted and reexamined. This interaction is done with satellite LANDSAT images, where the aim is to find and identify wheat crops, or enemy missile sites, to calculate the state of water resources, or the spread of urban development. The satellite data can be envisaged as a form of input for regional planning decisions. Images are analyzed and the data extracted from them is input to planning models that produce output which can be displayed superimposed on the original image. For example, the county boundaries, major highways and railroads can be superimposed on an aerial image of the urban areas of a larger state or country. In every county, numeric data derived from the image can be displayed, such as building density, percentage of vegetation, percentage of pollution in the air, etc. Then planning models can be used to help make decisions about major development plans such as new roads, new towns, locations of factories, dams, etc.

Optical Character Recognition

Another application is in optical character recognition with human assistance. Text is scanned and recognition performed whenever possible. Typically, 99 % recognition might be achieved, and unrecognized text can be displayed as an image together with surrounding recognized characters. From the context, an operator can identify an unrecognized character and type it on a keyboard. The character can be displayed under the image and if correct the operator continues to the next unrecognized character. If the operator makes a mistake or cannot recognize the image then scrolling and backup mechanics can be used to further aid context recognition. One can see the day when letters and memos .can be spoken into an office computer, recognized, displayed as an image, corrected and then distributed electronically. Such a useful device would need a signal (pattern) recognizer in a text display terminal. Input documents might be coded text or images from a scanner. Hardware developments and price reductions are beginning to make such text, graphics, image and speech systems possible.

APPLICATION EXAMPLES

Two applications using combinations of graphics and images on a color display system were reported previously [11]. A summary is presented here. A third example in which raster data (image) is used as a graphic primitive together with vectors and text is in our Picture Building System, also reported earlier [12] but not summarized here.

Map Editing

Map editing involves the entry, error detection, and error correction of digitized maps. We have developed a prototype system for this application and studied its use on graphics terminals with and without image mixing. A map-editing study was made with 12 maps containing from 1500 to more than 10,000 line segments each. The map editing was equally divided between terminals with and without image mixing. The map-editing software was the same for each terminal. Six users of approximately equal map-editing skills did the work; maps and terminals were randomly assigned to the users. The improvements in user productivity due to image mixing were:

Task	Improvement Factor	Base
map entry	4-7X	10 segments/hour
error detection	2-4X	no. of errors
error correction	2-10X	5 minutes/error

Maps are maintained by every utility, most government agencies, most companies dealing with natural resources (e.g., timber), and most companies with geographically dispersed customers or property. One user we worked with wanted to prepare 100 maps averaging about 3000 line segments each. Based on a factor of five productivity improvement over a base of 10 segments per hour, we estimate that the potential labor savings from using an image-mixing terminal rather than manual or non-image-mixing graphics to create these 100 maps is approximately 25,000 man hours. Another enterprise is trying to detect and correct errors in 250 maps containing over 50,000 line segments each, averaging about 10 % errors per map. Again using our productivity improvement estimates, the potentail saving with image mixing is between 250 and 400 man hours per map, or between 62,500 and 100,000 man hours.

Slide Making

In this application, the terminal hardware is used in a stand-alone mode and the complete application runs on the 'control' micro-processor. The user selects functions from a menu and colors from a pallette or paintbox. A joystick provides x, y input for drawing lines horizontally, vertically or at any angle, for drawing solid or open rectangles by specifying a diagonal, and for drawing solid or open circles by specifying a center and a point on the circumference. Text can be positioned and typed anywhere on the screen with or without a background surrounding color or on top of the existing picture. An image from a TV camera can be mixed in also, and used to form a composite picture. The menu and pallette can be turned off by a single keystroke. The displayed picture, exactly as constructed by the user can be photographed, and the resulting pictures or 35-mm slides can be used for technical presentations. Better quality output could be obtained by driving a higher-resolution precision film exposing device to produce a final slide or output. There are other useful features of the program: colors can be selected and changed at any time, for example, for every line while typing text, etc., a prompt system can be turned on/off to help a user learn how to operate the program, a picture can be stored on a cassette tape for later playback or mixing with subsequent pictures, a backtrack feature allows the last entry to be erased and some freehand sketching modes are provided. This is the second version of a program that was first written about four years ago on the first color terminal at approximately the same time that R. Schoup first developed his painting system (independently) [17]. Our first system was written to test our hardware and provide a simple demonstration capability, but the second system was built to assist the study of the human factors of man-machine interfaces and is intended to be very easy to use; like a copier almost. Cost analysis and productivity gains are given in Reference [11]. One can typically make slides for $ 3.00 each at a rate of about 10 per hour with little experience. This cost includes amortizing equipment over four years and using it continuously.

CONCLUSIONS

For a long time, image processing and computer graphics have been separate fields of endeavor. Individuals have been in one camp or the other, but rarely in both. Now common equipment for both graphics and image processing is becoming more widely available and workers are beginning to combine the advantages of both disciplines. As this offers advantages to both groups, the overlap is expected to increase in the future. In some applications, this is already happening as evidenced by some of the examples described.

REFERENCES

1. H.H. Poole, 'Fundamentals of Displays', Spartan Books, Washington D.C., 1966.

2. Beam penetration CRT called 'Penetron' used in CPS-8001 CRT monitor by CPS Inc. Sunnyvale, CA, 1973.

3. Genisco, Ramtek, Data-Disk, Aydin, Evans and Sutherland and Tektronix are example or industrial suppliers of color-display systems. There are many others.

4. Mitsubishi, Chuo-Musen, and Conrac make high-resolution CRT monitors. Matsushita makes high-resolution CRTs. There are probably others.

5. C.H. Herty and A. Manson, 'Color Plotter for Computer Graphics using electrically controlled Ink Jets', Proc. IFIP Congress '74, Stockholm 1974, published by North-Holland Publishing Co., pp. 85-88.

6. A. Rosenfeld, 'Picture Processing: 1977', Computer Graphics and Image Processing 7, April 2, 1978, pp. 211-242 (609 papers categorized).

7. U.W. Pooch, 'Computer Graphics, Interactive Techniques and Image Processing 1970-1975: a Bibliography', Computer, Vol. 9, No. 8, August 1976, pp. 46-64 (683 papers categorized).

8. Versatec, Gould, and IBM are example manufacturers of raster printers.

9. R. Williams, 'A Survey of Data Structures for Computer Graphics Systems Acm Computing Surveys', 3, March 1, 1971, pp. 1-21.

10. R.H. Ewald and R. Fryer, ' Final Report of the GSPC State-of-the-Art Subcommittee', ACM SIGGRAPH Computer Graphics, Vol, 12, Nos. 1 and 2, June 1978, pp. 14-169.

11. E.D. Carlson, G.M. Giddings and R. Williams, 'Multiple Colors and Image Mixing in Graphics Terminals', Proc. IFIP Congress '77, Toronto, August 1977, pp. 179-182, published by North-Holland Publishing Company.

12. D.L. Weller and R. Williams, 'Graphics and Relational Database Support for Problem Solving', ACM SIGGRAPH Computer Graphics, Vol. 10, No. 2, Summer 1976, pp. 184-189.

13. Bui Tuong Phong and F.C. Crow, 'Improved Rendition of Polygonal Models of Curved Surfaces', Proc. 2nd USA - Japan Conf. Tokyo, Japan, 1975.

14. F.C. Crow, 'Shadow Algorithms for Computer Graphics', ACM SIGGRAPH Computer Graphics, Vol. 11, No. 2, Summer 1977, Proc. of SIGGRAPH '77, San Jose, pp. 242-248.

15. C. Csuri, 'Computer Graphics and Art', Proc. IEEE Special Issue on Computer Graphics, Vol. 62, No. 4, April 1974, p. 503.

16. Adage, Evans and Sutherland, Vector-General and HUMRRO.

17. R.G. Schoup, 'Towards a Unified Approach to 2-D Picture Manipulation', ACM SIGGRAPH Computer Graphics, Vol. 11, No. 2, Summer 1977, p. 178.

18. Computer Graphics, Vol. 11, No. 3, Fall 1977, ACM SIGGRAPH, publication available from ACM HQ, 1133 Ave. of the Americas, New York, N.Y., U.S.A.

19. ACM Computing Surveys, Vol. 10, No. 4, Dec. 1978, the whole issue is on graphics standards.

MODEL-DRIVEN VISION FOR INDUSTRIAL AUTOMATION

L. Lieberman
IBM Thomas J. Watson Research Center
Yorktown Heights, New York
U. S. A.

INTRODUCTION

Automatic computer analysis of images for recognition, inspection, and verification has advanced to the stage where it is feasible for many industrial applications. Industry has a need for productivity enhancements and there is a growing acceptance of other forms of computer controlled automation (such as the industrial robot). The state-of-the-art in image processing techniques, the reduced cost and size and higher reliability of image sensors, and the trend toward low-priced, yet high-performance micro- and mini-processors seems to indicate that the time is "ripe" for the introduction of complex vision tasks to industry. However, the problem in implementing computer vision tasks in the past has been one of programmability, that is, the need for skilled programmers to generate the analysis program for each new task.

The industrial environment is favorable to vision programming since, in most cases, the task is known a priori, and knowledge of the (limited) domain can be incorporated in the program. Furthermore, the visual field is usually controllable; visual noise and clutter can be minimized and lighting can be engineered to make foreground/background separation and feature extraction easier. Also, the trend toward using Computer-Aided-Design (CAD) techniques has made parts databases available for applications programs. In the factory of the future the marriage of CAD and computer vision will change the way inspection and assembly are done.

There has been much recent research (Refs. 1-9) in industrial vision for use with robots or parts handling machines. Most of these have been concerned with quick methods of analysis. The problem of acquiring a "model" of a part for the recognition program to use is solved by one of four approaches (as described in (10)): Have the system acquire the model by showing it an instance of the object (and then analyzing the image of that instance), provide a computer-readable "drawing" of the object (the CAD approach), embed a description in a special-purpose program (the traditional method), or model the object in a simple, semi-natural language description.

The proliferation of CAD systems and the advantages of no image training and the availability of already created part models (i.e., in the modern, automated factory the parts will have been designed and entered into the computer from the beginning of the product cycle), make the CAD model-driven approach very attractive. Some of the most promising work in modeling for Computer-Aided Manufacturing include the BUILD program (11) and the Part and Assembly Description Language (12). A modeling program designed specifically for vision programming is GEOMED (13).

PROBLEM DEFINITION AND METHOD

The particular vision-manipulation problem addressed in this research is that of a general purpose parts feeder for use with a computer controlled industrial robot. The idea is to have a non-specific feeding mechanism that will 'drop' a part into an area in the robot's workstation and a video input system that will provide a digital image of the part. A program will then analyze the image to determine the part's 3-D orientation and position with enough precision to allow the manipulator's control system to properly grasp the part for its current task. The goals are to eliminate the high costs of special feeding mechanisms that need to be engineered for each part and to provide an automatically generated image analysis system.

The assumption is made that the field of view will contain only one object and that the background is uncluttered. The availability of lighting engineering (see Ref. 5) is also assumed so that objects are clearly outlined and easily separable from their background (a backlit platform is helpful for this). The 3-D recognition problem is reduced to a 2-D one. Thus, it is assumed that shape and topological features of regions are sufficient for determining the 3-D orientation (posssible future work involving a full 3-D model at image analysis time is discussed later).

The work reported below is the first step in an automated process for workpiece orientation recognition. The general method is this: Using a three-dimensional, computer-aided design system, form a realistic representation of the parts to be analyzed. A program will then automatically determine the stable orientations of the parts (within a rotation and a translation) on a horizontal plane. Another program will simulate the real camera's viewpoint and calculate views of the stable part orientations yielding the outlines (silhouettes). These silhouettes are in turn passed to a feature extraction program that determines values for a prototype feature vector to be used in real-time analysis and matching.

The object representation chosen is the polyhedron which, for a large class of industrial parts, provides adequate shape information. A powerful method for creating polyhedral models is described below. Programs to find stable orientations and object outlines were also completed. Work on the feature extraction and real-time analysis system are part of current research.

THE MODELING SYSTEM

In the course of research on a very high level programming language for computer controlled mechanical assembly (14), the need arose for a means to create realistic representations for scenes of objects. In particular, "world models" comprised of detailed descriptions of part shapes and spatial relationships are required for a planning compiler to solve such problems as generation of collision-free trajectories and calculation of feasible ways to grasp a part. A CAD system for creating such world models called the Geometric Design Processor (GDP) was developed. The object models created with the use of GDP are well suited to the vision problem outlined above. The system with examples of its use are described below.

GDP usage consists of two phases: a procedural description phase (15) in which the user writes a PL/I program that contains a structural composition of an object, and an interactive phase during which the user can view and modify the object model generated as a result of the model program's execution. A user-written object procedure is a template for how the object model is to be instantiated. Objects are "constructed" of volume pieces combined with the operations of union and difference. Perhaps the best way to describe this process is through an example:

```
INTERLOCK: PROCEDURE;

    /*  This is the object procedure that defines a
        typewriter part called an "interlock."
        A global body coordinate frame is known to
        this procedure and is manipulated through
        calls to routines like XYZTRAN and ZROT.
        The value of the global frame may be accessed
        through calls to STORE, and modified
        through RECALL   */

DCL (RAILFRAME(4,4),RAILFRAME2(4,4)) FLOAT;
CALL STORE(RAILFRAME);
```

```
/*  CALL SOLID produces a positive instance of the
       first argument, in this case the primitive
       CUBOID.  */

   CALL SOLID(CUBOID,'LEFT__BAR',.42,.90,.05);
   CALL XYZTRAN(0.3,.18,0.0);
   CALL SOLID(CYLNDR,'ROUND__PART',0.05,0.3,15.);

/*  CALL HOLE produces a negative instance, i.e.,
       the object is subtracted from what has already
       been formed.  */

   CALL HOLE(CYLNDR,'HOLE1',0.05,0.1,9.);

/*  Restore original coordinate frame for convenience
       of locating next component.   */

   CALL RECALL(RAILFRAME);
   CALL XTRAN(.65);
   CALL YROT(180.);
   CALL ZROT(-30.0);
   CALL ZTRAN(-.05);
   CALL SOLID(CUBOID,'DIAGONAL__SUPPORT',.65,.22,.05);

/* Several more volume components are created at this
       point in the procedure, but the code has been
       omitted for sake of brevity.  */
          .
          .
          .
/*  "MID__PART" is the user-defined object procedure
       that combines three bars to make the midsection
       of the interlock.   */

   CALL SOLID(MID__PART,'MID__SECTION');

   CALL RECALL (RAILFRAME2);
   CALL XYZTRAN(.98,.09,0.0);
   CALL SOLID(CUBOID,'MORE__OF__CENTER',.76,.25,.05);

/* "NLUG" is a user-defined object procedure.
       It is the notched, tab-like part of the interlock
       that protrudes from its midsection.   */

   CALL SOLID(NLUG,'NOTCHED__LUG');
```

This object procedure is typical in that it calls the procedures SOLID and HOLE to cause the addition or subtraction of an object. That "object" can be another object procedure or one of five primitives (cuboid, cylinder, cone, wedge, and hemisphere). An object procedure recieves a coordinate frame, or position cursor, as well as any formal parameters. The position cursor can be changed by calling system routines such as XROT (rotate about the X-axis). The parameters such as those in the statement CALL SOLID (CUBOID,...) are used to instantiate the generic description in the procedure (e.g. a CUBOID is defined by three length values).

The activation sequence of the calls to SOLID and HOLE contained in object procedures is a tree structure. The system automatically generates a data structure analagous to this tree, where each node represents an object component and branches indicate the "part-of" relationship. That is, an object component at a node in the tree is a part of the object represented at its parent (predecessor) node. Each of the object nodes has several attributes. For the purposes of vision processing, the most important attribute is the polyhedral equivalent of the object. For primitive nodes (leaves), the system directly instantiates the polyhedron. For any higher-level node, it applies a general polyhedra merging algorithm to the polyhedra of the subtrees. This algorithm can form the union or difference of any pair of polyhedra. The operation is chosen according to the "polarities" (SOLID or HOLE) of the nodes taken pairwise, left to right. Figure 1 shows the primitive component pieces of the interlock part defined above and the resulting, merged object.

The interactive phase of GDP allows the user to view at a display console the structures previously created. One can modify objects and store the results on file for future use. Thus, a "scene" generation facility is achieved for both complex vision and manipulation tasks. The interactive processor has many useful commands: Objects may be viewed from any user-definable 3-D viewpoint, and can be repositioned and reoriented, then "merged" to form new objects. Features of objects, specifically, faces, edges, and vertices, can be identified by the user through pointing with the display cursor. These features can be named and referred to later. Objects can be positioned symbolically with the POSITION command: "POSITION interlock SUCH THAT interlock.side CONTACTS table.top," where interlock and table are objects and side and top are faces.

GDP was augmented with a set of programs for solving the stable orientation problem and generating 2-D outlines (silhouettes). The program STABLE takes as input a polyhedron and returns a list of support planes defining "stable" orientations of that polyhedron. The method used first calculates the convex hull of the object. CONVEX does this by applying a gift wrapping algorithm to the polyhedron's vertices. Figure 2 shows the hull computed for the interlock of Figure 1 The "gift wrapping" algorithm finds an extreme edge and plane, then

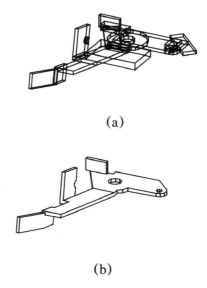

(a)

(b)

Figure 1. (a) The interlock part component pieces and (b) the
result of 'merging'.

proceeds to "wrap" around the edge to find the nearest set of points
forming a new planar face. The convex set in this plane is found and
interior points discarded. The edges formed by this set are added to the
list of edges for the hull and wrapping proceeds for any "unused" edges
in the list. This is quite efficient since the marking of used and discarded vertices eliminates unneeded calculations as the process continues.

The next step in the STABLE program is to find the object center
of mass with the program CMASS. This program assumes constant
density of object material. It integrates the volume formed by the
"prism" under each polyhedral face edge-loop. These component prisms
are positively weighted for the loops containing surface material and
negatively weighted for the loops around holes. The centroids of the

Figure 2. The convex hull of the interlock part.

prisms are combined to give the object's center of mass. STABLE then projects the center of mass onto each of the convex hull faces, thus hypothesizing that the face is a stable support plane. If the projected point lies within the bounds of the face polygon, then it is tentatively declared a stable plane. Next, a test is made for the degree of stability by calculating the energy needed to "tip" the object by rotating over the edge in the support plane nearest the projected center of mass. If the energy is large enough, i.e., exceeds a user-set threshold, then the convex face is a support plane.

All the faces are checked and those passing the tests are ranked in decreasing order of stability (i.e., "tipping" energy) and returned in a list to the calling program. Figure 3 shows the stable positions found for the interlock. Figure 4 shows another typewriter part in its calculated stable orientations.

Both STABLE and CONVEX are called from the interactive GDP. The next step in the model driven vision process, also called from GDP, is to find the silhouette of the object. The object model is manipulated via the POSITION command for any of the stable support faces of the convex hull (note that the SUCH THAT clause will now involve the hull faces as variables to control the position of the original object). The user sets the imaginary camera viewpoint to simulate the real camera's placement in the workstation and commands the system to GAZE at the object (this essentially aims the principal ray of the camera system at the centroid of the object model). The program OUTLINE then calculates the outer boundary of the object's projection into the display plane (the zoom factor is set by the SCALE command). The vector list is stored in a file and optionally displayed. Figures 5 and 6 show the outlines for each of the two parts in their stable orientations as viewed from directly "above" the block platform.

FUTURE WORK AND SUMMARY

A method for modeling industrial parts and programs to compute stable orientations and their views were described. Work is now continuing on the problem of feature selection for discrimination of stable positions and for determination of the object's rotation in the plane. It is hoped that a method for automatically choosing a combination of low-level features (from a menu of such feature extractors) will be forthcoming.

Another approach is to not discard the 3-D model at recognition time as described above where 2-D features are extracted as a prototype vector, but to make use of the models in the real-time recognition program. The techniques of parametric correpondence and chamfer match-

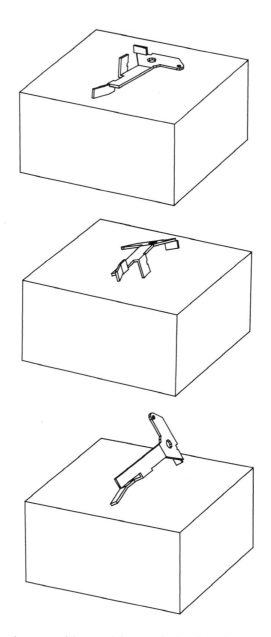

Figure 3. The three stable positions of the interlock part as calculated and drawn by GDP.

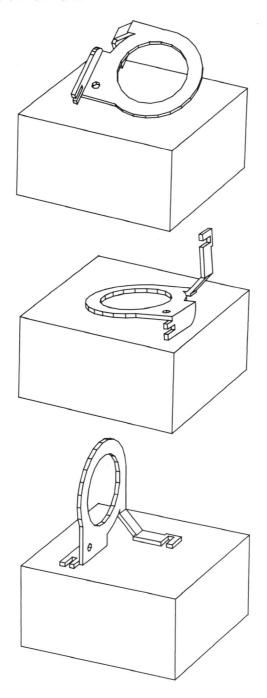

Figure 4. A set of stable orientations for another typewriter part.

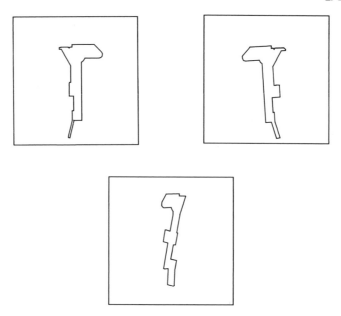

Figure 5. The silhouettes calculated for each of the orientations of Figure 3, viewed from directly overhead.

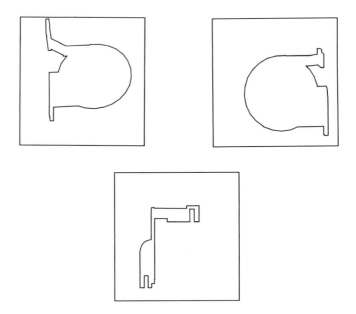

Figure 6. The silhouettes calculated for each of the orientations of Figure 4, viewed from directly overhead.

ing (16) could be used to vary the position and orientation of the 3-D model, project it into the image plane, then measure the closeness of match between feature points in the real sampled image and the projected model image. This sequence is iterated by changing the parameters of the model's coordinate frame to minimize the feature difference function ("hill climbing"). This approach (and most others employing 3-D models at recognition time), however, could require much computer time.

ACKNOWLEDGEMENTS

The author wishes to thank M. A. Wesley, M. A. Lavin, and T. Lozano-Perez for their efforts in designing and programming GDP, D. D. Grossman for providing the basis for procedural representations of objects, R. Evans for his contributions to the design of the polyhedral merging algortihm, and P. M. Will for inspiring all of our work.

REFERENCES

1. W. A. Perkins, "A model-based vision system for industrial parts," *IEEE Transactions on Computers, Vol. C-27, No. 2*, pp. 126-143, Feb. 1978.

2. M. L. Baird, "Image segmentation technique for locating automotive parts on belt conveyors, in *Proc. of the 5th International Joint Conference on Artificial Intelligence*, pp. 694-695, Cambridge, Mass., Aug. 1977.

3. R. C. Bolles, "Verification vision for programmable assembly," in *Proc. of the 5th International Joint Conference on Artificial Intelligence*, pp. 569-575, Cambridge, Mass., Aug. 1977.

4. J. Bretschi, "A microprocessor controlled visual sensor for industrial robots, *Industrial Robot*, pp. 167-172, Dec. 1976.

5. G. J. Agin, "An experimental vision system for industrial application," *Proc. of the 5th Int. Symposium on Industrial Robots*, Chicago, pp. 135-148, 1975.

References 6 to 9 appeared in the *Proc. of the 8th International Symposium on Industrial Robots*, Stuttgart, May 1978.

6. J-D. Dessimoz, "Visual identification and location in a multi-object environment by contour tracking and curvature description," in the *8th ISIR*, pp. 764-777.

7. A. Niemi, "Object recognition and handling in an industrial robot system with vision, in the *8th ISIR*, pp. 744-755.

8. J. Birk, *et. al.*, Acquiring workpieces: three approaches using vision, in the *8th ISIR*, pp. 724-733.

9. A. Pugh, *et. al.*, Versatile parts feeding package incorporating sensory feedback, in the *8th ISIR*, pp. 206-217.

10. S. W. Holland, "An approach to programmable computer vision for robotics, *General Motors Research Report GMR-2519*, Warren, Michigan, Aug. 1977.

11. I. C. Braid, *Designing with Volumes*, Cambridge, England: Cantab Press, 1974.

12. "An Introduction to PADL," Production Automation Project Technical Memorandum 22, Univ. of Rochester, Rochester, New York, Dec. 1974.

13. B. G. Baumgart, Geometric modelling for computer vision, Ph.D. Thesis, Stanford University, Stanford Artificial Intelligence Laboratory Memo AIM-249, Oct. 1974.

14. L. I. Lieberman and M. A. Wesley, "AUTOPASS: An automatic programming system for computer controlled mechanical assembly, *IBM Journal of Research and Development, Vol. 21, No. 4*, pp. 321-333, July 1977.

15. D. D. Grossman, "Procedural representation of three dimensional objects, *IBM Journ. of Research and Development, Vol. 20, No. 6*, pp. 582-589, Nov. 1976.

16. H. G. Barrow, *et. al.*, Parametric correspondence and chamfer matching: two new techniques for image matching, in *Proc. of the 5th International Joint Conference on Artificial Intelligence*, pp. 659-663, Cambridge, Mass., Aug. 1977.

IMPLEMENTATION

DISTRIBUTED IMAGE PROCESSING

W. Giloi
Techn. Universitaet Berlin
Berlin
Federal Republic of Germany

Abstract

Image processing is usually approached as the problem of applying numerical algorithms to digitized gray-scale images and may be very expensive in terms of memory space and computation time. Therefore, it is worthwhile to look for more economical solutions. An analysis of typical image-processing tasks reveals that the usual costly approach can often be avoided or, in certain cases, may not even be appropriate. These considerations led us to the design and development of a special "Picture Processing Display Terminal" (PPDT) which has been devised for performing most of the required pre-processing and feature-extraction steps at the periphery of the computer at a small fraction of the costs that arise when the same processing is carried out in the computer. Hence the input to the computer already consists of binary representations of significant features, thus considerably reducing the amount of data to be further processed. Furthermore, each preprocessing step, which may be performed either with the man in the loop or automatically, is made visible in order to provide the feedback and the opportunity to evaluate the performance of the system and correct mistakes. The PPDT is part of a system for "distributed image processing". The considerations leading to the design of such a system are discussed in this paper, as well as its possibilities and limitations.

Workload Distribution in Image Processing

The realm of image processing may be classified into two major subareas, namely, image enhancement and analysis. In image enhancement, the result is an image of improved quality. In image analysis, on the other hand, the result is a certain understanding or evaluation of an image. This paper deals with image analysis.

Image analysis is a two-step procedure. The first step performs the task of feature extraction, and in the second step numerical or logical operations are performed on the extracted features. Thus, only the feature-extraction part — which may include a number of steps for preprocessing and picture segmentation — deals with gray-scale *images*, whereas the image understanding or evaluation algorithms deal with *features*, that is, do not operate on the original picture elements.

In the approach presently dominating in image analysis, the images to be analyzed are first digitized; preprocessing and feature-extraction procedures are subsequently performed on the matrix of the gray-scale values obtained. Digitalization of an image is accomplished by scanning the image, sampling the analog gray-scale values, and converting the discrete values obtained in this manner into digital numbers. However, it is not absolutely mandatory to proceed in such a way. On the contrary, it may be much more economical to perform certain preprocessing or even feature-extracting functions on the analog gray-scale values prior to the conversion into digital numbers.

The feature-extraction process itself is typically a multistep procedure. In the first step, the image is subjected to certain pre-processing operations such as, for example, noise filtering, contrast enhancement and edge sharpening. Subsequently, the preprocessed image is converted into a binary pattern by thresholding. Thresholding may be considered as a segmentation technique which produces certain features given by the obtained binary patterns [3]. In most cases, however, further segmentation procedures are applied to the binary pattern (e.g., skeletonization, thinning, shrinking, border following, region growing, distance measurement, etc.) in order to eventually obtain features which are as characteristic and nonredundant as possible. These features are the objects of measurement, classification, or scene-analysis algorithms.

Usually, all these steps are performed on digitized gray level pictures by applying a sequence of numerical operations. This approach is so overly expensive that in many cases it prohibits any practical use of image processing, e.g., in the delivery of health care. This seems to us to be the main reason why, despite so many research efforts, image processing has not found its routine application in the clinical practice. However, as our brief analysis of the whole process of image processing (from which we exclude image enhancement as an end in itself) reveals, the overly expensive "brute force" approach, which starts with a digital gray-level representation, is very poorly matched to the hierarchical nature of image analysis as a multistep process starting with applying *simple operations* (point transformations or local operations) *on a vast amount of data* (the picture elements) and ending up with applying *highly sophisticated procedures* (for measurement, classification, or scene analysis) *on a greatly reduced amount of data* (representing the extracted features). The simple operations performed on the picture elements do not require the computing power of a general-purpose computer. They can be accomplished by an imaginative use of hardware. The more complicated procedures, used toward the end of the process, do require the computing power of a general-purpose computer. On the other hand, they are performed only after a tremendous data reduction has been achieved by the preceding steps. Thus, the economy of image processing can be greatly enhanced in many applications if the borderline between hardware preprocessing and computer processing is shifted from the very beginning of the multistage process (where it is now) to an appropriate later stage in the process.

For the applications which we have in mind, we suggest placing this borderline between the thresholding step and the subsequent more complicated segmentation procedures. Hence, such a "matched" image processing system needs to encompass hardware for edge detection in the original image, for instance, by a computation of the gradient, thresholding, sharpening of the thresholded image by thinning and noise filtering. Such a hardware component is the "Picture Processing Display Terminal" which was designed and developed with a grant from the Federal Republic of Germany at the Heinrich-Hertz-Institute in Berlin, Germany.

The Picture Processing Display Terminal (PPDT)

The PPDT was originally designed for interactive or "semiautomatic" use in image processing. To this end, a TV raster display was developed in which graphical representations can be generated on the basis of a 386 × 256 dot matrix. The gray-level information for each dot is stored in an associated one-bit memory cell (so-called "bit mapping"). Each dot can assume one of the two possible levels, "blank" or "bright". Hence, any arbitrary binary picture can be displayed with a resolution given by the 384 × 256 raster. Besides such binary pictures, text and special symbols may be displayed. The text capacity is 48 × 21 = 1008 characters per frame.

Graphical objects can be generated by a computer program as well as by the user. Thus, the user can enter a graphical dialog with the system. To this end, he is provided with a keyboard for typing in alphanumerical characters and special symbols as well as with a light pen for free-hand drawing. As the main purpose of the display originally was to function as a tool for interactive feature extraction performed on photographic images, the system allows the superposition of a video image and graphical objects generated by aid of the dot matrix on the same screen. Hence, a user can look

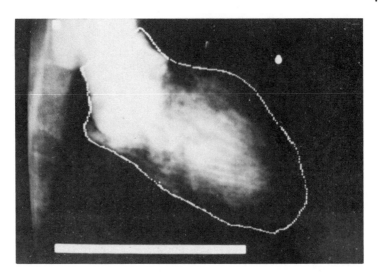

Figure 1. Boundary line of left ventricle manually marked by lightpen.

at a photographic image and then use the lightpen for tracing edges,
borderlines or other landmarks, contours or areas, etc. An example
is presented in Figure 1. It shows the ventriculogram of the opaci-
fied left ventricle. The boundary was outlined by a radiologist using
the lightpen. All the objects thus created, representing features
in the image, are stored in the dot-matrix memory and are displayed
in superposition with the video image. To the host computer of the
PPDT, these objects are binary pictures which were created and identified by a human operator with his superior cognitive faculties.

Subsequently, it became the design objective to extend the
capabilities of this very useful device so that it could function
as the input and preprocessing device in automated image processing.
To this end, five special hardware units were added which accomplish
the following tasks: contrast enhancement, calculation of the gradient, thresholding, thinning and noise filtering. The solutions
which were found shall be discussed in more detail.
    A. *Calculation of the Gradient by Differentiation*. The video
signal is supplied by a TV camera, a video disc, or any other source
of video images. It represents the gray-level values of the image
along the scanning lines of the TV pattern. An example is Figure 2
which shows a standard coronary angiogram.
Differentiating this signal yields the horizontal component of the
image gradient. Since the video signal is a continuous function of
time, differentiation requires only a very simple analog network.

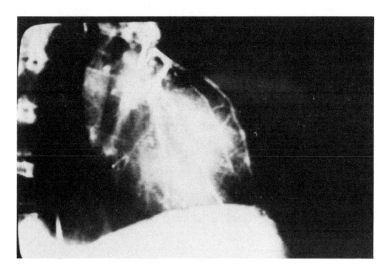

Figure 2.   Coronary angiogram; Original video image.

The vertical component of the gradient, however, cannot be obtained
in such a simple manner since it requires the calculation of the
difference between the gray-level values of all pairs of corresponding points of two adjacent line scans. To this end, the gray-
level values of all points of a line scan must be stored for the
duration of one scan line so that they can be subtracted from the
respective values of the following scan line (which is the one
currently represented by the video signal). Therefore a delay line
with the delay of one scanning period must be provided. Digital
delay lines are available in the form of shift registers. Their use,
however, necessitates a preceding fast analog-to-digital conversion
as well as a digital subtraction (which is more expensive than an
analog subtraction network). On the other hand, analog delay lines
calibrated for exactly one line scan are a component of any PAL color television receiver. These delay lines are mass produced and
are inexpensive. Their technical maturity makes them a rather
reliable device. It was found that the analog solution is not only
the less expensive but also the less critical approach with respect
to distortion and noise. Figure 3 is the differentiated image of the
angiogram shown in Figure 2. Figure 3 more clearly shows the main
coronary arteries and the location of the catheter.

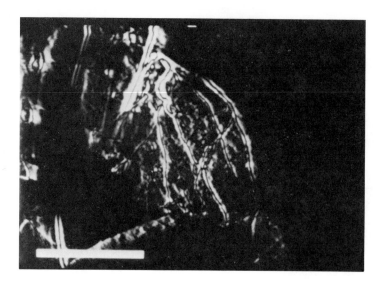

Figure 3.   Coronary angiogram; differentiated image.

B. Thresholding. As the video signal or the differentiated video signal is still an analog function, an analog comparator can be
used for thresholding. The threshold can be set by the host computer
via a digital-to-analog converter. Actually, the system contains
two such comparators and converters. By connecting the Boolean outputs of both comparators by EXCLUSIVE OR, the subset of all picture
elements of equal gray level is converted into a binary pattern
("isophots"). The binary pattern obtained by thresholding is stored
in the dot matrix memory, so that it can be displayed (in superposition with the original gray-scale image) and/or fetched by the
host computer for further processing. Figure 4 represents the binary picture which resulted from thresholding the differentiated i-
mage of Figure 3 with a particular threshold setting. Different threshold settings permit enhancement and suppression of details in the
image and can therefore serve presegmentation purposes.

Figure 4. Coronary angiogram; differentiated and thresholded image.

   C. Thinning. High-contrast black and white edges appear relatively rarely in radiographic images. Typical edges vary gradually
from low to high gray values and vice versa. Therefore, the gradient shows wide maxima rather than a sharp and needle-like appearance. The thresholded gradient of such moderately sharp edges consists of extended clusters rather than of sharp lines. For edge detection, it therefore becomes necessary to sharpen the binary image
by a thinning technique. Therefore, the differentiated analog signal
is thresholded separately in horizontal and vertical direction. A
digital logic replaces all horizontal strings of "1's" obtained by
the thresholding by the left-most "1" and all vertical strings of
"1's" by the top "1". The thinning logic decides that a point be-
longs to a (thinned) edge when either an edge in the horizontal
direction or in the vertical direction or in both directions has
been encountered.

   The differentiation of an image not only enhances edges and
contours but also the noise in the image. Therefore, it is advisable
to rid the binary pictures of noise before entering them into the
dedicated computer, the more so as this can be done without affecting the useful information. To that end, a local operation is performed to remove isolated dots and fill isolated "holes", that is,
reduce noise. An example for a thinned and noise-filtered image is
presented in Figure 5.

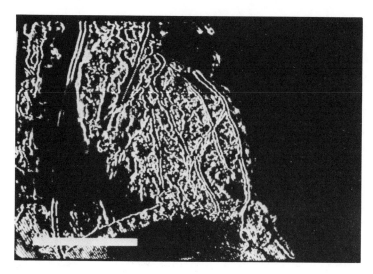

Figure 5. Coronary angiogram after differentiation, thresholding, thinning, and noise filtering.

Preprocessing Potential of the PPDT

Basically, the image processing potential of the PPDT is determined by the possibility of thresholding and thinning contrast enhanced gray scale images or differentiated (edge enhanced) images. The possibility of controlling the setting of a threshold by the host computer enables the host computer to obtain a histogram of the gray level values of an image by performing a series of thresholdings with varying settings. Likewise, a "stack" of binary pictures, extracted with different threshold settings from the same gray scale image, can be obtained. Binary pictures, or stacks of pictures, are the objects of a variety of algorithms for fusing, shrinking, border following, calculation of a distance measurement, calculation of the skeleton, etc. The preprocessing and thresholding, including thinning and noise filtering, takes the time of one TV picture frame, i.e., about 20 milliseconds. Hence, the transfer of the dot matrix memory content (96 K bits) or the processing time in the host computer, whichever is greater, is really the factor limiting the rate at which gray-scale images can be processed.

The Image-Processing System

The PPDT replaces the more expensive gray-level scanner/digitizers used in conventional image processing systems. Moreover, it alleviates dramatically the computational burden placed on the host computer, for the latter has to deal only with binary patterns. The above-mentioned algorithms are all of a simple nature, so that a computer of moderate power can be employed. On the other hand, the PPDT and its host must be close enough to permit a direct data transport via the DMA channel of the computer. The common voice grade communication links cannot be used, for even at a relatively high rate of 2400 baud it would take 40 seconds to transmit the dot matrix memory content from the PPDT to the host. Therefore, the PPDT is directly connected to a small-scale computer which, in turn, is connected via a high-speed link to a large computer system. Hence, a distributed system was obtained that is perfectly matched to the decreasing amount of data and the increasing complexity of operations which are characteristics of the sequence of steps in image processing. The PPDT preprocesses half-tone images into binary pictures, and the "dedicated" small-scale computer performs the logical or arithmetical operations on these binary patterns, resulting in condensed features. The host system then will accomodate programs for performing quantitative measurements, classification, or scene analysis. The computing power of the host system allows practically any desired complexity of operations (as may occur in statistical classifiers, picture grammar approaches, etc.). Costs are reduced by orders of magnitude as compared with conventional image processing.

As an example for the role of the dedicated computer (e.g., a minicomputer) in the system, we discuss the border reconstruction procedure, in which the dedicated computer is instrumental. One or more portions of an object boundary within an image may not exhibit a large enough gray-scale gradient to be extracted directly by the PPDT along every scan line that intersects the boundary. This results in an extracted edge which is not continuous but rather is broken by gaps and missing segments. Further, the noise present in an image may introduce false points into the characteristic function (binary picture) that represents the extracted edge information. Hence, the binary edge picture developed by the PPDT must be postprocessed by the dedicated computer in order to remove unwanted noise points as well as insert missing edge sections so that the edge can be properly reconstructed.

The noise points generated by the PPDT can be reduced by applying a higher threshold level to the gradient signal. However, as the noise disappears with an increase in the threshold, so do portions of the boundary which are characterized by a small gradient. Therefore, a threshold level must be found that passes most of boundary information and yet does not allow too many noise points to be inserted into the binary picture. Experimentation has shown that the optimal threshold level differs from one part of an image to another, and therefore an algorithm had to be developed to allow the automatic selection of an optimal threshold for each of several subregions of a total image.

Once a proper thresholding operation has been accomplished, the number of noise points as well as the number and length of the border gaps will be small enough to allow effective post-processing of the binary picture. The post-processing needed to close the border gaps and eliminate the noise points can usually be accomplished by local operations on the binary picture. This reconstruction of the boundary is accomplished in two major steps.

In the first step, each of the "1's" in the binary picture are expanded by inserting 1's in either the N_4 or N_8 neighbors of the existing 1's. If the N_4 neighborhood is used, then the expansion is called a diamond expansion since a single "1" element grows into a diamond shaped region. Likewise, if the N_8 neighbors are set to 1's, then a square expansion is accomplished. The diamond expansion grows fastest in the horizontal and vertical directions whereas the square expansion grows fastest along the diagonals of the image. If the diamond and square expansion are applied alternately, then the growing occurs more evenly in all directions, resulting in an octagon shaped region growing from a isolated point. The octagon expansion is usually used because it eliminates the directional bias inherent in the diamond and square expansions, although it is slightly slower than the other two methods. The main feature of the expansion which makes it useful to the reconstruction process is that it fuses together disconnected segments of the boundary if the expansion is iterated a sufficient number of times.

In addition to the desirable effect of fusing together disconnected segments of the boundary, the expansion has the undesirable effect of widening the boundary into a thick region. A second post-processing step is necessary to thin the boundary so that the exact edge position is clearly indicated. The thinning algorithm [4] removes edge points from a thick region repeatedly until the region is reduced to a connected line of points which is chosen to be either a simple four-path or a simple eight-path. One important feature of the thinning algorithm is that it maintains the topology of the binary picture, so that an edge which has been connected by the expansion process outlined above remains connected after the thinning process. Another important feature of the thinning algorithm is that it thins a region toward the centerline of the region.

Once the thinning process is complete, and the edge has been reconstructed, the application-dependent analysis can be easily accomplished by using a local border following algorithm. Thus the data are extracted on which the subsequent evaluation and classification procedures performed by the host system are based. Figure 6 illustrates the effect of the border reconstruction procedure on a coronary angiogram.

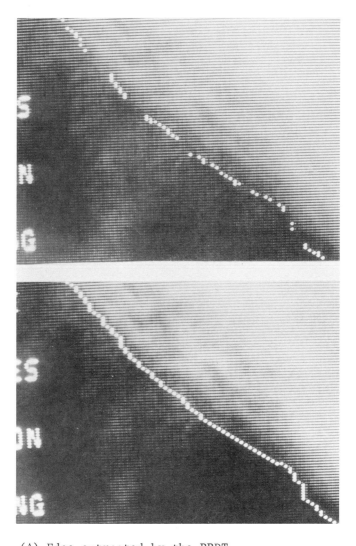

Figure 6. (A) Edge extracted by the PPDT.
           (B) The same edge after application of the border recon-
               struction program.

Man-Machine Interaction

One of the very helpful features of the PPDT is that, during the research phase in which certain concepts will have to be tried, every preprocessing step can be monitored and judged by the investigator since the content of the dot matrix memory can always be displayed, superimposed to the original image. Thus, the extracted edges or borderlines can be viewed as being drawn over the original image. Furthermore, the user can take the lightpen and aid the system, if need be. For example, the PPDT permits the user to mark with the lightpen an "area of interest" and ask the system then to concentrate only on that area. Thus, a step-by-step improvement of the automated image preprocessing algorithms can be achieved interactively.

The demand for such interactive capabilities stems from the fact that, despite the tremendous practical importance and despite the large amount of theoretical work that has already been invested in this field, automatic picture information processing and pattern recognition by computers is still far from being a solved problem. It is commonly anticipated that it will take many more years before more effective solutions will be found to solve this problem satisfactorily and economically in terms of computer utilization.

Experiences with the System and Future Development

To evaluate the capabilities of our system, we developed an application package for the automated recognition of the heart contour and the evaluation of the heart volume over a series of frames of a cine-angiographic film. In general, this project proved to be feasible under the proviso that the recognition of disturbing artefacts (diaphragm, ribs) was accomplished in a semi-automatic fashion. To this end, the first frame of the series of pictures was manually processed by the radiologist, simply by tracing the heart contour, as seen by him/her,with the lightpen. From then on, a series of heart contour extractions was automatically performed on approximately 80-100 frames. The data obtained were used to evaluate the dynamics of the heart wall movement, providing information of the highest diagnostic value.

Whereas the problem of distinguishing an object from artefacts is a pattern recognition rather than an image processing problem, we came to recognize that the optimal threshold setting in the PPDT is probably the biggest problem in the feature-extraction portion of image analysis. It may be very hard to find criteria that can be used to adapt the threshold setting to a given problem in an optimal or at least reasonable way, and this problem calls for heuristic solutions developed as the result of elaborate experimentation. Figure 7 block-diagrams the total "distributed" image-analysis system discussed here.

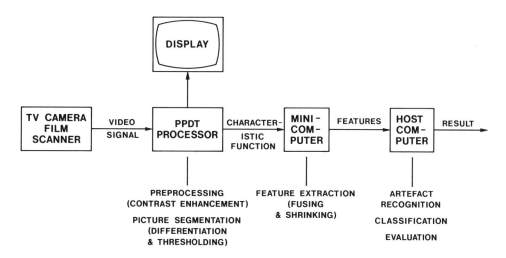

Figure 7. Hierarchical system for distributed image processing.

However useful the PPDT, as described in this paper, may be as an experimental device, its routine application in image analysis would be handicapped by the fact that the hardware of the system is special-purpose and, therefore, somehow difficult to replicate. The "hybrid" circuits used in the system give rise to certain problems (the PAL delay line, for instance, turned out to be a performance-limiting factor) and, in general, exhibit the lesser reliability of such circuits as compared with digital circuitry.

What we would want to use is strictly digital hardware — except for the analog-to-digital converter needed to convert the video signal into digital form — which consists solely of "off-the-shelf" components. On the other hand, we would not like to sacrifice too much of the speed advantage which the hybrid solution provides.

What are the requirements imposed by this application, and to what extent can these requirements be met by standard minicomputer hardware?
(A) *Memory requirements.* We assume (1) a maximum frame size of 1024 × 1024 pels, (2) a 1-byte gray-level representation for each pel, (3) the inappropriateness of "in-place computation". With these assumptions the memory capacity must be >2 M-bytes. Standard memory capacity of next-generation minicomputers will be 1 M-byte or more, so that this requirement can be satisfied.

(B) *Desired processing speed.* The processing speed should be high
enough to render memory bandwidth, the only speed-limiting factor. In next-generation minicomputers, the maximum data flow rate
will be four to eight M-bytes/s, leading to a desired execution
time for any operation on the picture matrix in the order of magnitude of one μsec. However, this is not achievable with conventional hardware and the "von Neumann fashion of operation."

The best we can do is to use modern, high-performance minicomputers that have a fast writable control store. On such a machine,
all operations required to access data (indexing, address calculation),
as well as the simple local operations that have to be performed in
the early stages of image analysis can be implemented by firmware,
that is, put into microcode. This results in a speed-up factor —
in comparison to software routines — somewhere in a range from 10
to 50. Of this speed enhancement, at least a factor of 5 can be attributed to the fact that, in a firmware implementation, most instructions
are microinstructions rather than instructions of the conventional
machine language. Therefore, they are fetched from the writable control store rather than from main memory. At present, the cycle time
of writable control store is approximately five times shorter than
the cycle time of semiconductor main memory. Additionally, a speed-
up factor between 2 and 10 can be attributed to the fact that, in
contrast to conventional machine language coding, microcoding allows
for a certain degree of parallel execution (what this degree actually is depends on the nature of the operations and the machine architecture).
With these measures, the execution time of simple image processing
operations may be estimated as follows.
1. *Fetch and store.* Does not require the execution of a machine instruction and, therefore, is small (~1 μsec.) compared to the execution time of the operation to be performed on the fetched pel.
2. *Contrast enhancement (logarithmic transform).* Point operation,
   performed by a table-lookup in a ROM (required ROM capacity: 2^8 =
   256 bytes). Hence, this operation is roughly matched in speed to
   memory bandwidth.
3. *Differentiation or thresholding.* Involves arithmetical and relational operations. Order of magnitude of execution time: ~10
   μsec/pel (mismatch by an order of magnitude).
We recognize that, on the average, the execution speed is 1 order of
magnitude slower than what memory bandwidth would permit, albeit it
is at least 1 order of magnitude faster than what can be accomplished by machine language routines. The processing of an image consisting of 10^5 pels in the same way as is performed by our special-
Hardware PPDT would take several seconds. This is two orders of magnitude slower than the PPDT (20 msec./frame). However, one has to
bear in mind that the minicomputer which performs the subsequent
feature extraction (cf. Fig. 7) cannot keep up with such a high
speed anyway.

As a result of this discussion, we may state that firmware implementation of the various steps of image processing (excluding
global operations such as Fourier transform, etc.) on a cascade of
minicomputers will yield a system that is still matched to the hierarchical nature of image analysis. Although considerably slower than
special-purpose hardware, such a system would be faster by at least
an order of magnitude than any conventional general-purpose computer.
On the other hand, the use of standard digital hardware would offer
the highest-possible degree of economy. Nothing would be gained by
performing all steps of the hierarchical image analysis procedure
on one and the same large-scale computer, as the firmware implementation of the preprocessing operations is, in any case, much faster
than any software implementation on a large-scale machine.

Acknowledgements

The author is gratefully indebted to Mr. Steven L. Savitt and
Mr. Claus E. Liedtke for their contributions to this work. He also
wants to express his great appreciation to the German Federal
Government, Ministry of Research and Technology, to Dr. Frank
Verbrugge, Director of University Computing Services, and to the
Graduate School, both of the University of Minnesota, for supporting
this research project. Dr. R. Moore of the Department of Radiology
must be given thanks for many helpful discussions and for providing
the imagery with which the system was tested.

References

[1]   W. K. Giloi and C. E. Liedtke: "An Image Processing System
        for Radiological Application", Proc. of the San Diego Bio-
        medical Symposium 1976, Vol. 15.

[2]   W. K. Giloi: "Moving the Hardware-Software Boundary up in
        Hierarchical Multi-Stage Image Analysis", Proc. of the
        Milwaukee Symposium on Automatic Computation and Control
        (1976), 439-443.

[3]   A. Rosenfeld: "Picture Processing by Computer", Academic
        Press, New York, 1968.

[4]   S. Yokoi et al.: "Topological Properties in Digitized
        Binary Pictures", Systems, Computers, Controls, Vol. 4,
        No. 6 (1973), 32-40.

PARALLEL PROCESSORS FOR DIGITAL IMAGE PROCESSING

M. Duff
University College
London
England

INTRODUCTION

Parallel processing is an ill-defined concept. To some, parallelism implies little more than duplication of arithmetic units, or some sort of provision for simultaneous operation of various parts of an otherwise conventional computer architecture. For others, parallelism is represented by 'pipelining', in which the data streams through successive computational units, each carrying out its own particular operation on the data. Again, a parallel system can be one in which the basic machine instructions involve simultaneous access to a substantial subset of the total data to be processed.

However, in this paper, a discussion will be presented of what may be regarded as an extreme form of parallel architecture, in which a necessarily small, special-purpose processor is dedicated to each element of the two-dimensional data field being processed. Computers structured in this way are classified as 'Single-Instruction Multiple-Data Stream', using the useful classification scheme devised by Flynn[1].

Data sets representing images exhibit certain properties which suggest what should be the characteristics of computers designed to process them. In the first place, it is clear that many useful operations on images can be conceived either as sequential operations or as parallel operations. For example, if an image is to be shifted a little to the right, then the elements composing the image may be shifted one by one or all together; the results of both processes are identical. Similarly, in a black and white picture in which the figure is white and the background black,

edges elements of the white figure can be determined serially by
following round the black/white borders, or in parallel by identifying all white elements adjacent to black elements. The conceptually parallel algorithms can also be implemented in a serial mode
by visiting sequentially each data element and by noting sequentially the states of neighbouring elements surrounding it.

Suppose an image is represented by a rectangular array of
square elements Pij, where (i,j) are the array coordinates of a
picture element ('pixel'), and Pij is the grey-level intensity
of the pixel at (i,j). It is convenient to let P range from 0
(for black elements) to 1(for white elements). Usually P will be
quantised into a finite number of levels, 64 being both convenient
and typical. The set of elements in a neighbourhood surrounding
the element Pij can be written {Pij}S, where S specifies the size
of the surrounding neighbourhood.

A parallel process can be defined by writing:

$$P_{ij}^* = F(P_{ij}, \{P_{ij}\}S) \text{ for all } (i,j).$$

Where Pij* is the new value of each element Pij following the
operation F. In a serial process:

$$P_{ij}^* = F(P_{ij}, \{P_{ij}\}_{S1}, \{P_{ij}\}^*_{S2}),$$

where {Pij}S are the elements of {Pij}S as yet unaffected by the
operation F, and where {Pij}*S2 are elements of {Pij}S which have
already experienced the operation F. To clarify these definitions,
suppose the operation F on a binary (black/white) image consists
of changing from white to black each element whose left neighbour
is black. If F is applied in parallel, then the effect will be to
spread black regions to the right by one element. If, however,
the picture is subjected to a raster scan in which F is applied
sequentially, then all black regions will extend to the right edge
of the picture, which is a very different result. Thus the set
{Pij}S becomes {Pij}*S2 and {Pij}S1 is empty for all (i,j) except
at the left side of the array. Note, in passing, that the neighbourhood S needs additional definition at the edges of the array.

It can be shown that there is a duality between serial and
parallel operations on images. Any parallel process can be implemented as a sequence of serial operations and vice versa. It is
well known that parallel algorithms can be programmed onto serial
computers; it is perhaps less generally appreciated that serial
algorithms can be programmed onto general-purpose parallel computers.
This point is discussed in some detail by the author in [2].

The second important factor to be taken into consideration when designing computers for image processing is that data representing an image is structured and that this structure is usually most significant over small localities. This rather imprecise statement is nevertheless of great importance to computer architects since it implies that much can be found out about an image by making measurements on the small groups of pixels surrounding each pixel. In terms of the notation introduced above, the neighbourhood set S can be the immediate neighbourhood set N (the eight elements surrounding each pixel in a square array). Furthermore, by iteratively applying F in a modified form, the effective range of S can be extended to any required distance, since more distant elements in S will be neighbours of neighbours etc.

In summary, the special nature of data representing an image points to the suitability of a computer composed of an array of identical processors, one for each pixel, and each receiving data both from its 'own' pixel and from the pixel's immediate neighbours. The processors are arranged so as to operate in parallel and all perform the same operation at the same instant. The output from the array is a new set of pixels representing a 'processed' image. In the next section, the realisation of these ideas in a cellular logic image processor, CLIP4, will be described.

CLIP4

CLIP4 is the most recent of a series of cellular logic arrays which have been constructed in the Image Processing Group of the Department of Physics and Astronomy at University College London. UCPRI was a special purpose, multilevel, fixed function array designed for a specific 'task' (vertex detection) in the analysis of bubble chamber photographs[3]. Subsequently, a study of variable function (programmable) arrays was carried out, based on a diode array described in [4]. In the same paper, the first of the CLIP series of arrays is also reported. CLIP2 is described in [5] and CLIP3 is reported and analysed in [2]. All these arrays, with the possible exception of the first, can be seen as design studies leading to the design now being realised as CLIP4. CLIP3 was first operated in 1973 so there has been plenty of time to allow user experience to be acquired. At the time of writing (September 1978), most of CLIP4 has been constructed and first samples of an integrated circuit comprising eight processors have been tested. A further redesign cycle will be necessary in order to bring up to specification the electrical performance characteristics of the integrated circuit.

The concept of an array of processors for image-processing applications is not new; probably UNGER (6) was one of the first to discuss a practical design for such an array. The historical

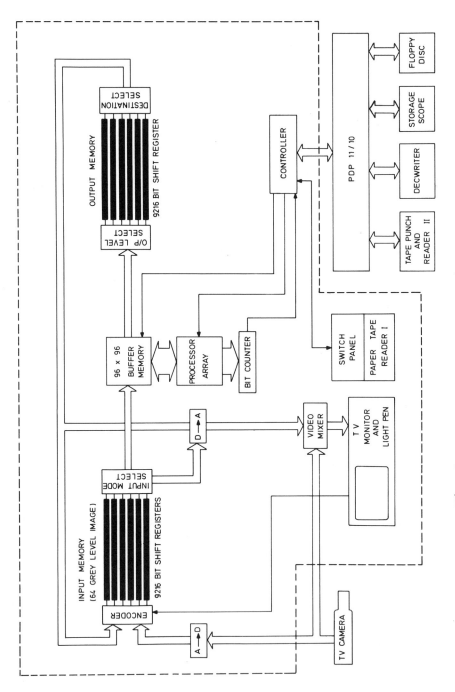

Fig. 1.   The CLIP4 Image Processing System.

development of these ideas is reviewed in [2] and [5]. The CLIP project has paid particular attention to the economics of array processing. Since processors are assigned to every pixel, and since pictures may be expected to require some thousands of pixels in order to provide adequate resolution over an acceptable area, then it is obvious that cost will limit the complexity of the processor for the vast majority of users. On the other hand, if the logical capability of each processor is too small, then, at the best, many functions will require long strings of oversimplified parallel operations and, at the worst, some functions may be impossible to implement. The compromise between cost and efficiency is hard to arrive at and will not stay optimised against a background of developing technology in the semiconductor industry.

The CLIP4 system is shown in Fig. 1 and contains 9,216 processors arranged in either rectangular or hexagonal configuration as a 96 by 96 pixel array (selectable under program control). The image is obtained from a television video signal, utilising the central third of the middle 96 scan lines. The analogue signal is digitised to 6 bits (64 grey levels) and stored external to the array. A 6-bit output picture store is also provided, with D-A conversion permitting the display of a grey-level processed image. Television scan rate A-D and D-A conversion, and also thresholding, is achieved. A special-purpose controller and instruction store operates the array and its I/O facilities. Additionally, a PDP11 interface is provided to permit more flexible control and interfacing to peripherals.

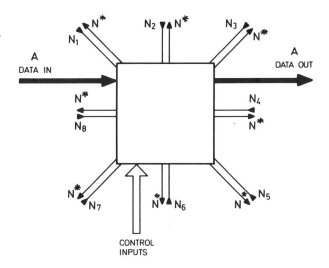

Fig. 2.   Interconnections between the Processors in the Array.

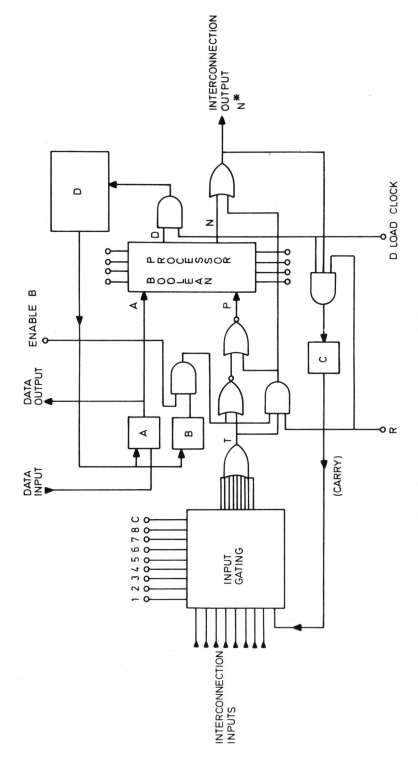

Fig. 3. Logic of the CLIP4 Processor.

Within the array, every processor includes 32 bits of storage, D_0 to D_{31}, and 3 bits of special-purpose buffer storage, A, B and C. Connections between processors are shown in Fig. 2. The complete CLIP4 processor logic is shown in Fig. 3.

Each of the incoming connections, N_1 to N_8, can be individually gated under program control. The operation of the circuit can be best understood by considering Figs. 4 to 9 which each show selected parts of the logic. In every part, the heart of the processor is the two input, two output boolean processor.

The two independent boolean functions, B_N and B_D, are selected by two sets of four control lines. There are 16 possibilities for each function. The so-called 'process' instruction specifies the two boolean functions, the enabled interconnection inputs, the array configuration (square or hexagonal), the input to be made to interconnection inputs at the array edges (1 or 0), and the state of the B, C and R enable lines (see Fig.1). Other instructions are 'load' instructions which load the A and B buffers from the D memories, and various types of conditional and unconditional branches. Instructions operating on the 14 general-purpose registers in the controller are also available. Similarly I/O instructions operate on both the array and the 6 bit memories. Programming techniques for the array are outlined in several publications (see, for example, (2)).

Referring to Fig. 4, the simplest array operation is to form a boolean function of two binary patterns, loaded into buffers A and B. Next in complexity are the immediate neighbour operations (see Fig. 5) in which the second boolean output N (which is identical to N* here) is transmitted to neighbours. For example, if $B_N = \bar{A}$ and $B_D = P.A$, then the processor will output as a pattern of

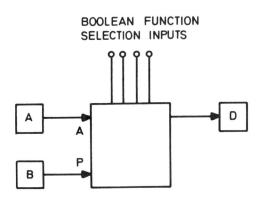

Fig. 4.   Processor Logic for Simple Boolean Operations.

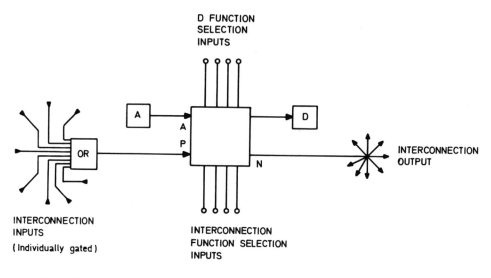

Fig. 5.   Processor Logic for Immediate Neighbour Operations.

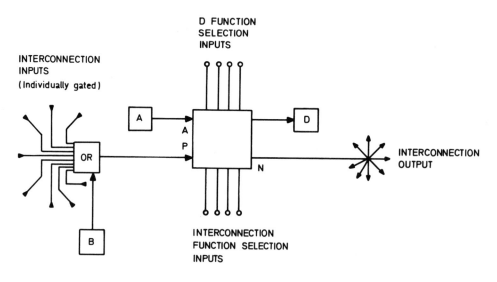

Fig. 6.   Processor Logic for Labelled Propagation.

1's the edges of a figure composed of 1's on a background of 0's.
In Fig. 6 a further sophistication is included: 1 elements of a
second binary pattern loaded into B will 'inject' an interconnection input into the array. As an example of the use of this technique,
consider a pattern representing several unconnected white objects
(composed of 1's) on a black background. Suppose that one element
of an object to be selected is known. This appears as a pattern
comprising a single 1 element in a field of 0's. The first pattern
is loaded into A and the second (the 'label') into B. The functions
selected are $B_N = P.A$, $B_D = P.A$. An interconnection signal origi-
nates at the label in B and flows through all elements of the
required object. The boolean function B_D then selects those processors whose A input is 1 (i.e., part of an object) and whose P
input is 1 (i.e., one of the processors connected to the label).
Thus the required object is output into the selected D address as
a pattern of 1's.

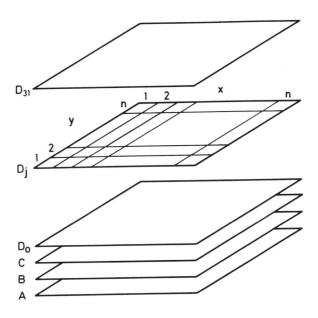

Fig. 7. Organisation of Array Storage.

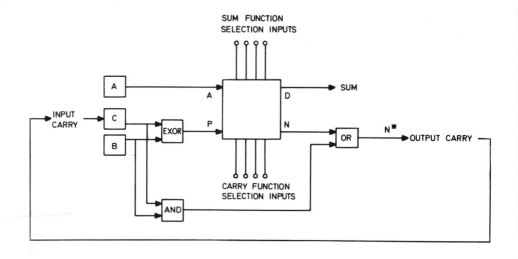

Fig. 8.   Processor Logic for Bit-Plane Stack Arithmetic.

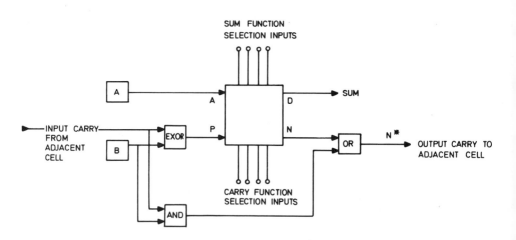

Fig. 9.   Processor Logic for Binary Column Arithmetic.

The arithmetic capability of CLIP4 is illustrated in Figs. 7, 8 and 9. Fig. 7 shows the organisation of the array storage. A field of 9,216, 6-bit numbers can be stored in, say, D_0 to D_5. This is usually called a 'bit-plane stack' or 'bit stack'. Two bit stacks can be added, subtracted or compared by loading corresponding bit planes into A and B and by selecting B_D to give the bit sum and B_N to give the carry (see Fig. 8). One operation is required for each bit-plane addition. The carry bit plane is stored in C. The processor behaves as a full 'bit-plane' adder and can be used to achieve more complex arithmetic such as multiplication, division etc.

In another mode (see Fig. 9) binary numbers are stored as columns in a single bit plane. Two such bit planes, each containing 96, 96-bit numbers can be compounded arithmetically by means of a sequence of 4 parallel operations.

OPERATING SPEEDS

Process times for images are not easy to calculate for either serial or parallel systems and it is therefore difficult to evaluate the advantages, in terms of operating speeds, of using parallel processors. Simplistically, it might be guessed that, other things being equal, an N by N array of processors would be N^2 times quicker than the single processor of a serial machine. Unfortunately, 'other things' are not equal. First, the parallel processors have direct access to neighbouring processors and will experience a further advantage. Second, local storage in the array eliminates time spent on addressing during fetching and storing. Third, some of the available process instructions are particularly powerful for image processing. But in the other direction, it must be realised that the parallel processor has to be cheap, and this implies a slower semiconductor technology with a less sophisticated structure. Attempts have been made to evaluate CLIP systems with respect to conventional computers and with application to a range of image processing tasks [7]. For a 96 by 96 processor array, speed gains range from 10 to over 10^6 for the tasks studied. In practice, CLIP3 has been shown to operate as much as 10^6 times faster than an IBM360/65 for particular image processing operations.

Care must be taken not to exaggerate claims for complete image-processing systems based on parallel processing. For many of the simpler processes, process times are dominated by I/O times for the vast amounts of data which constitute the images being processed. In such instances, pipelined systems will often prove to be equally, if not more, cost effective and fast. However, it is confidently predicted that with the advent of larger and larger integrated circuits and with the increasing skills of parallel processing programmers, image-processing operations will become very much more complex, so that the advantages of an array of processors will become more apparent. In CLIP4, simple boolean operations take rather less than 10 microsec, with an additional 1.2 microsec for propagation of interconnection signals through each processor beyond nearest neighbours. The 96 by 96 pixel image can be entered into the array as a binary picture (by thresholding) and some 1,500 parallel operations performed during each television scan cycle; grey-level processing requires two cycles for the same number of operations. At these speeds, real-time image processing becomes a feasible proposition.

ACKNOWLEDGMENTS

The author is grateful to the U.K. Science Research Council and to University College London for their financial support of the CLIP program, and acknowledges the contributions of ideas and effort to this project by all the members of the Image Processing Group.

REFERENCES

1. FLYNN, M.J. "Some computer organisations and their effectiveness", IEEE Trans. on Comp., 948, (Sept. 1972).
2. DUFF, M.J.B. "Parallel processing techniques". Chap. VI of "Pattern Recognition Ideas in Practice", ed. B.G. Batchelor, Plenum (1978).
3. DUFF, M.J.B., JONES B.M. and TOWNSEND, L.J. "Parallel processing pattern recognition system UCPR1", Nucl. Instrum. Meth., 52, 284 (1967).
4. DUFF, M.J.B. "Cellular logic and its significance in pattern recognition", AGARD Conf. Proc. No. 94 on Artificial Intelligence, 25-1 (1971).
5. DUFF, M.J.B., WATSON, D.M., FOUNTAIN, T.J. and SHAW, G.K. "A cellular logic array for image processing", Patt. Recog., 5, 229, (1973).
6. UNGER, S.H. "A computer orientated toward spatial problems", Proc. IRE 46, 1744, (1958).
7. CORDELLA, L., DUFF, M.J.B. and LEVIALDI, S. "Comparing sequential and parallel processing of pictures", Proc. 3rd Int. Joint Conf. on Pattern Recog., San Diego, 703 (Nov. 1976).

LARGE-SCALE VECTOR/ARRAY PROCESSORS

G. Paul
IBM Thomas J. Watson Research Center
Yorktown Heights, New York
U. S. A.

INTRODUCTION

The general subject of array processing [1 to 7] has been a topic of considerable study since the 1950s as both system designers and users have sought to improve performance and price/performance. The results of these studies have led researchers and designers to explore numerous design alternatives and means of achieving greater concurrency and hence performance. The multiplicity of these approaches has grown so diverse as to completely obfuscate the meaning of the term "array processor." The term is now commonly used in reference to machine designs including arrays of processing elements (both SIMD, single-instruction multiple-data stream; and MIMD, multiple-instruction multiple-data stream designs), as well as pipelined vector instruction processors, associative processors, algorithmic processors and other designs optimized to processing arrays of data efficiently. Table I below presents a brief taxonomy with examples of the various types of "array processors" available commercially today. A list of supplemental references is appended to papers on machines not described herein.

In this paper we shall focus our attention on only one of these types of array processing systems: vector instruction processors. We shall compare the representative machines in this category and summarize the relative strengths and weaknesses of their approaches.

Before we delve into specific comparisons, let us first consider common approaches utilized to enhance performance in large-scale computing systems. In general, design techniques to improve performance are directed to either increase the level of concurrency of instruction execution of the processor or to enhance the effective bandwidth of data flow to match the execution bandwidth of the various functional units in the processor.

I. Vector Instruction Processors
    A. Memory to memory
        CDC STring ARray Processor, STAR-100
        Texas Instruments Advanced Scientific Computer, ASC
    B. Register to register
        Cray Research CRAY I

II. Parallel Processor Ensembles
    Burroughs ILLIAC
    Burroughs Parallel Element Processing Ensemble, PEPE
    ICL Distributed Array Processor, DAP

III. Associative "Array" Processors
    Goodyear Aerospace Corporation STARAN
    Sanders Associates OMEN

IV. Algorithmic "Array" Processors
    IBM 2938 Array Processor
    IBM 3838 Array Processor
    IBM Master Data Processor
    CDC Matrix Array Processor
    Univac Array Processor

V. Algorithmic Mini-"Array" Processors
    Floating-Point Systems Array Transform Processor
    Datawest Array Transform Processor
    CSP, Inc. Macro Arithmetic Processor

Table I.   Partial Classification of Array Processor Architectures and Machine Organizations

Approaches to enhance concurrency of execution within the instruction decode unit include pipelining, look-ahead/look-aside logic, loop entrapment, instruction and operand prefetch, branch prediction, etc. Within the execution function units, approaches include the utilization of multiple (parallel or independent) functional units, pipelining, common data bus forwarding, chaining, and so on.

Approaches to enhance effective bandwidth of data flow to match execution bandwidth include pipelining of memory (interleaving), high-speed cache memory, provision of additional high-speed register files and generally broadening of machine data paths to permit greater data access per unit time.

Array processing techniques generally provide an architectural framework in which these approaches may be used effectively in the internal machine organization or design.

PIPELINING

The pipelining technique implies the segmentation or partitioning of a process into subprocesses or levels which may be carried out independently, but not necessarily concomitantly, by independent units.   These subprocess units are then arranged in sequence such that successive executions of the process may be staged through the sequence of subprocesses thereby achieving concurrency of operation.   The instruction decode process is commonly segmented into the sequence of subprocesses as follows:

i)   IF -- instruction fetch,
ii)   ID -- instruction decode,
iii)   AG -- address generation, and
iv)   OF -- operand fetch;

the decoded instruction and operands are then forwarded to the appropriate function unit for execution.

For the sequence of instructions (A, B, C, ...), this process may be carried out concurrently as illustrated below:

Subprocesses

OF			A	B	C	D	E
AG		A	B	C	D	E	
ID	A	B	C	D	E		
IF	A	B	C	D	E			

Time

As indicated above pipelining is also commonly used within execution function units. For example floating-point multiply may be segmented as follows:

i)   Exponent add,
ii)   prenormalization/exponent adjust,
iii)   multiplication of mantissas, and
iv)   postnormalization/exponent adjust.

Through pipelining, if we assume S segments or levels are utilized in the pipeline and each segment requires time T, then N operations may be performed in the time, T_N:

$$T_N = [S + (N - 1)] * T,$$

however, each individual operation requires $S*T = T_S$ time units. This time, T_S, is commonly called the startup time for the pipeline. Refer to [1] for a more complete survey of pipelining.

Vector processing, as we shall see, is particularly amenable to pipelining techniques.

VECTOR PROCESSING

As is readily apparent, the technique of pipelining is only effective when N, the number of operations to be executed, is large with respect to S. The efficiency of the simple pipeline described above is generally defined by the following equation:

$$\text{Efficiency} = N / [S + (N - 1)],$$

where as N approaches infinity the efficiency approaches unity.

The primary advantage of a vector instruction processor is that a single vector instruction specifies a large number of identical operations are to be performed. Hence the duty cycle and efficiency of the pipeline is high, and the ancillary machine resources may be scheduled in advance. Furthermore, vector instructions also provide other advantages:

Vector instructions allow the programmer to directly convey to the computer explicit information about the program structure. This information is often lost or hidden in programs written for a conventional scalar processor, and must be recovered or reconstructed implicitly by the scalar processor through the use of stochastically adaptive mechanisms such as cache memories, branch history tables and/or other means of logical analysis of the program instruction and data flow. This information can also be utilized by the hardware to provide for systematic data prefetch and to generally schedule the required system resources in an orderly manner.

Since a single vector instruction schedules the machine activities for a relatively large number of cycles with respect to a conventional scalar instruction and implicitly specifies a loop, vector instructions generally reduce and simplify the processing load on the instruction decode unit. Vector instructions reduce the required rate of instruction decode, the number of instructions (and hence memory bandwidth required to supply instructions) and in particular they reduce the number of load, store, index and branch instructions. The reduction in branch instructions is especially significant as branch instructions generally cause delays and require special treatment such as alternate branch path fetch and decoding, branch prediction and other expensive logic to maintain high instruction throughput.

Vector instructions are also useful to the scientific programmer directly because they allow him to compactly specify his problem to the computer and reflect the underlying mathematical formulation and linear nature of most scientific applications.

Finally, with appropriate syntactic extensions and high-level language support [8-10], the same structural information conveyed to the computer by vector instructions can be utilized to increase the efficiency and optimization capabilities of the language processors (e.g., compilers or translators).

VECTOR INSTRUCTION PROCESSORS

The vector instruction processors first became available commercially in the early 1970s. Included in this class of processors are the Control Data Corporation STAR-100 (STring ARray processor), the Texas Instruments ASC (Advanced Scientific Computer) and the Cray Research Corporation CRAY I. Each of these systems feature a fully-integrated vector instruction set in the central processing unit along with a general-purpose scalar instruction set.

These machines are predominantly used for large-scale scientific computation such as numerical weather forecasting, climatological modeling, structural analysis, seismic data processing, simulation of nuclear reactors, aerodynamic simulation and analysis of weapon effects, etc.

We shall consider each of these machines individually as they each represent unique machine organization approaches.

CDC STAR-100

The STAR-100 [11] is designed such that its vector instruction set executes memory to memory. The central processing unit, see Figure I below, is partitionable into four basic functions: i) storage access control, ii) stream control, iii) floating-point processing units and iv) string processing unit. Each of these functions operate asynchronously and in parallel.

The storage access control unit controls the read and write buses shared by the streaming units and I/O subsystem, and contains the virtual memory addressing mechanism of the system.

The stream control unit is the basic control mechanism for the system. It includes the basic instruction decode mechanism, generates memory references for both instructions and operands, schedules and buffers operand flow, provides control signals to the execution units and performs simple logical and arithmetic operations. The stream unit contains a high-speed register file of two hundred and fifty-six 64-bit registers. An exit force instruction is provided to force a swap of the register contents

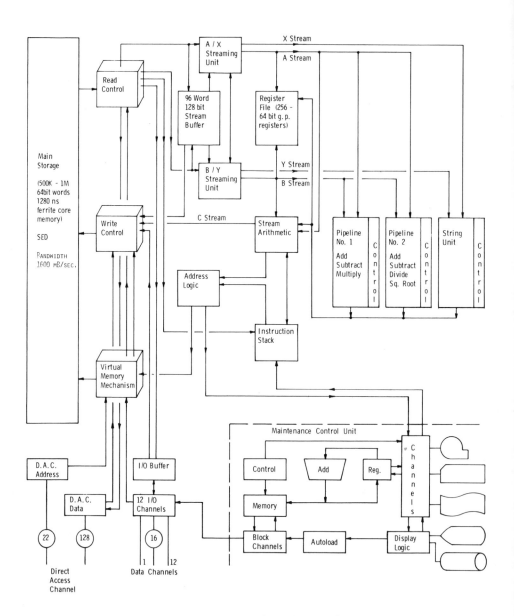

Figure I. The CDC STAR-100 Machine Organization

with data in memory to facilitate task switching in a multiprogramming environment. The instruction decode unit contains an instruction stack which may contain up to sixty-four 128-bit instructions. The stack mechanism allows for look-ahead of eight to sixteen instructions, and executions of loops of up to forty-eight instructions without memory access.

The two floating-point arithmetic units are pipelined as shown in Figure II below. Pipeline Unit No. 1 performs 32 and 64-bit floating-point addition, subtraction and multiplication. Pipeline Unit No. 2 performs 32 and 64-bit floating-point addition and subtraction, division and square root. Each pipeline is capable asymptotically of performing fifty million 32-bit or twenty-five million 64-bit additions or subtractions per second. Pipeline Unit No. 1 is capable of performing up to one hundred million 32-bit or twenty-five million 64-bit multiplications per second. Pipeline Unit No. 2 is capable of performing up to fifty million 32-bit or twelve and one half million 64-bit divisions per second.

The string processing unit is used to process either decimal or binary strings, performs the Boolean, binary and addressing arithmetic functions and processes control vectors during floating-point streaming operations. (The role of control vectors in STAR vector instructions will be discussed below.)

Although the STAR-100 features a forty nanosecond logic clock period, it has a relatively slow ferrite core memory (twelve hundred and eighty nanosecond). In order to maintain a high memory supply data rate, the memory is highly interleaved (thirty-two way) and has eight 32-byte parallel read and eight 32-byte parallel write paths to the independent memory read and write control units. Four of these paths may be active each cycle providing an asymptotic bandwidth of 1600 megabytes per second. Two of these buses are used to stream input operands to the execution units; the third to store results and the fourth is shared between I/O memory references and control vector references.

In order to utilize this potential bandwidth as fully as possible, the vector instructions in STAR are architected to operate only with vector operands whose elements are stored in contiguous words of memory. Since FORTRAN employs column-major storage ordering, this implies that only columns may be used as vector instruction operands directly; row vectors may not. As a means of programming around this architected column affinity, the STAR-100 provides a matrix transpose instruction. However, there are many algorithms where explicit transposition is costly, and frequently both row and column vector access are needed concomitantly as in matrix decomposition algorithms and in the ADI (alternating directions implicit) algorithm which is commonly used in the solution of systems of parabolic partial differential equations.

Although the STAR-100 vector instructions have a biased affinity to column vectors, the STAR instruction repertoire provides perhaps the richest functional capability of the three machines considered herein. This richness is particularly evident in the STAR's utilization of bit vectors to control storage of computed results in vector operations and as order vectors to specify positional significance in sparse vector instructions.

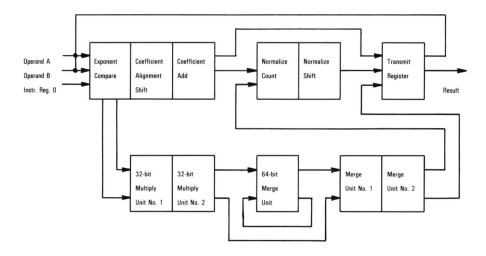

Floating-Point Pipeline Unit No. 1

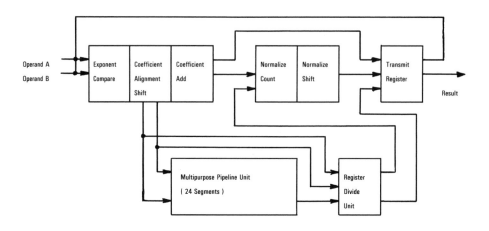

Floating-Point Pipeline Unit No. 2

Figure II. CDC STAR-100 Floating-Point Units

The role of control vectors in the STAR is illustrated by the following example. Consider the general vector instruction format shown below:

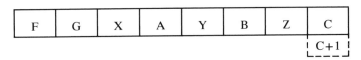

where

F	Function operation code,
G	Subfunction operation code,
X	Address of register containing offset address of operand A,
A	Address of register containing operand A field length and base address,
Y	Address of register containing offset address of operand B,
B	Address of register containing operand B field length and base address,
Z	Address of register containing control vector Z base address,
C	Address of register containing result vector C field length and base address,
C+1	Implicit register containing offset for C and Z.

If after resolution of the offset address displacements from the respective base addresses the operand vectors A, B and C are

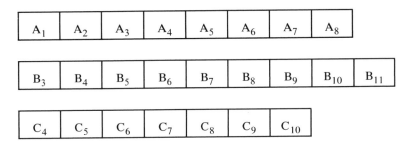

and the control vector Z has the value

0	1	1	0	1	0	0

then the result of the vector multiply instruction will be

C_4	A_2*B_4	A_3*B_5	C_7	A_5*B_7	C_9	C_{10}

Note, the control vector Z controls only the storage of results, the products (A_1*B_3), (A_4*B_6), etc., are generated but simply not stored to memory.

The role of the bit sequences as order vectors in the STAR sparse vector instructions is illustrated below. Consider the sparse vector instruction format

F	G	X	A	Y	B	Z	C

where

F	Function operation code,
G	Subfunction operation code,
X	Address of register containing A order vector length and base address,
A	Address of register containing base address of operand A,
Y	Address of register containing B order vector length and base address,
B	Address of register containing base address of operand B,
Z	Address of register containing result order vector length and base address,
C	Address of register containing length and base address of result vector C.

If the operand vectors A and B, and their respective order vectors have the following values:

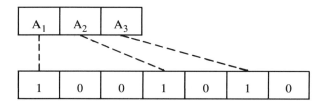

and

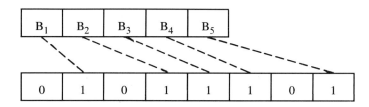

then the sparse add operation causes the following results to be generated for C and its order vector:

A_1	B_1	A_2+B_2	B_3	A_3+B_4	B_5		
1	1	0	1	1	1	0	1

In addition to the general and sparse vector instructions illustrated above, the STAR also has vector macro instructions such as sum, inner product and max/min reduction, polynomial evaluation, etc., as well as string instructions, etc.

The STAR-100 is supported by a general multiprogramming virtual operating system and language processors. The STAR FORTRAN [12] compiler features vector syntactic extensions to FORTRAN as well as automatic vectorization capability.

The Texas Instruments ASC

The TI-ASC [13] like the CDC STAR-100 is designed such that its vector instruction set executes memory to memory. However, the TI-ASC has a unique control structure for specification of vector instructions. The ASC central processing unit, Figure III below, consists of three types of units: i) the instruction processing unit (IPU), ii) the memory buffer units (MBUs) and iii) the arithmetic units (AUs). A particular system configuration may consist of one or two IPUs and one to four AUs or pipes, each with a corresponding MBU.

The instruction processing unit, IPU, is the basic control mechanism for the ASC system. It performs instruction fetch/decode, target register selection, address resolution, operand fetch/store, branch determination, register conflict and scalar hazard resolution, and provides control signals to the MBUs and AUs.

The memory buffer units performs the load/store function to memory and buffers data to and from the arithmetic units. Each memory buffer unit contains forty-eight 32-bit buffer registers. These registers are divided into six stacks of eight registers each with two stacks allocated to each of the two input operand streams X and Y, and the remaining two stacks are used to buffer the output stream Z. Refer to Figure IV below.

Each arithmetic unit is a general-purpose pipeline which can be configured under micro-control to perform floating-point or other arithmetic functions.

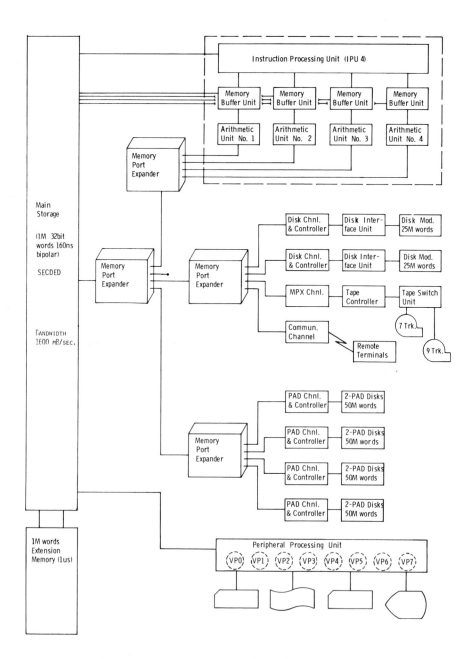

Figure III. The TI-ASC Machine Organization

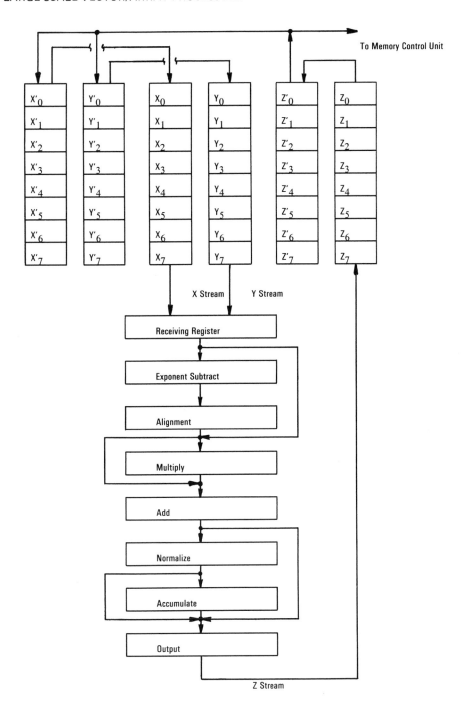

Figure IV. The TI–ASC Pipeline

The concept of a vector instruction in the ASC is considerably different from that of the STAR-100. The ASC contains a hardware macro facility which may execute a specified operation over a one, two or three dimensional address space. Vector operations are specified via a hardware vector parameter file or VPF contained in the IPU. The vector parameter file consists of eight 32-bit registers. The registers are partitioned into subfields as shown in Figure V below. The subfields specify the required information to describe the operation to be performed and the control information: starting addresses, increments, counts, etc., specifying a triply-nested DO-LOOP structure over which the operation is to be executed.

The innermost loop of this structure is called the self-loop. The outermost loop is called the outer-loop; the intermediate loop is called the inner-loop. The inner-loop and outer-loop may be incremented by arbitrary values through each of the three operand arrays A, B and C. However, the innermost loop, the self-loop may only increment in positive or negative unit steps. This limitation forces a (FORTRAN) column affinity in the self-loop analogous to that seen in the STAR-100 vector instructions. The column affinity of the ASC, however, is not as rigid as that of the STAR-100, since the self-loop may in effect be suppressed.

Once declared, the vector instruction specified by the VPF may be executed by a single execute instruction. Other instructions are provided to load and execute, store and otherwise manipulate a vector parameter file description.

Use of the vector parameter file reduces the vector setup time for the user since he may reuse parameter files once established. The VPF also simplifies the instruction decode process and provides fixed control logic paths to the read-only-memory (ROM) control unit which configures the flow of the ASC pipelines for the specific operation to be executed.

Like the STAR-100, the ASC has, in addition to vector primitive instructions, more complex instructions for inner product reduction, max/min reduction, etc. The instruction repertoire of the ASC, however, is not as diverse as that of the STAR. In particular, the ASC provides limited sparse vector support.

The ASC is supported by a general multiprogramming operating system and has language processors for FORTRAN, COBOL and PASCAL. The FORTRAN [14] compiler also provides a set of vector syntactic extensions as well as automatic vectorization capability.

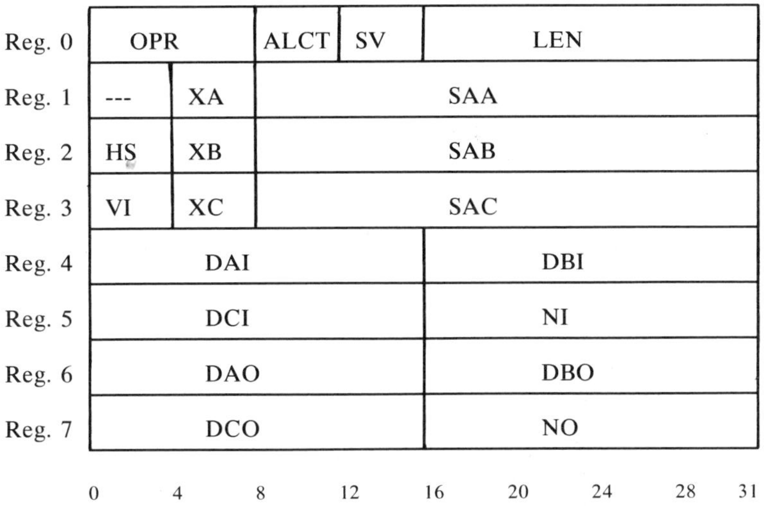

Reg. 0	OPR		ALCT	SV	LEN	
Reg. 1	---	XA	SAA			
Reg. 2	HS	XB	SAB			
Reg. 3	VI	XC	SAC			
Reg. 4	DAI			DBI		
Reg. 5	DCI			NI		
Reg. 6	DAO			DBO		
Reg. 7	DCO			NO		

```
0     4     8     12    16    20    24    28   31
```

where

OPR	Type of vector operation,
ALCT	Arithmetic or logical comparison condition,
SV	Single-valued vector,
LEN	Vector length (self-loop count),
XA	Starting address index of vector A,
SAA	Starting address of vector A or immediate A,
HS	Right or left halfword starting address,
XB	Starting address index of vector B,
SAB	Starting address of vector B or immediate B,
VI	Self-loop increment direction,
XC	Starting address index of vector C,
SAC	Starting address of vector C,
DAI	Inner loop increment for vector A address,
DBI	Inner loop increment for vector B address,
DCI	Inner loop increment for vector C address,
NI	Inner loop count,
DAO	Outer loop increment for vector A address,
DBO	Outer loop increment for vector B address,
DCO	Outer loop increment for vector C address,
NO	Outer loop count.

Figure V. The TI–ASC Vector Parameter File

The CRAY I

The CRAY I [15] is designed such that its vector instruction set executes register to register. The CRAY I features a set of eight vector registers, each vector register consisting of sixty-four 64-bit scalar elements. The CRAY I also features twelve independent functional units:

Functional Unit	Segmentation Length
Address add	2
Address multiply	6
Scalar integer add	3
Scalar logical	1
Scalar shift	2-3
Scalar leading zero/population count	3-4
Vector integer add	3
Vector logical	2
Vector shift	4
Floating-point add	6
Floating-point multiply	7
Floating-point reciprocal	14

Each of these functional units are fully segmented (pipelined) as shown to produce a result every twelve and a half nanoseconds. The data flow of the CRAY I is shown in Figure VI below.

The vector instruction repertoire of the CRAY I consists only of primitive vector/vector and scalar/vector arithmetic and logic instructions which operate register to register. The contents of the vector-length register (VL) control the number of operand pairs participating in a vector instruction. Vector operations over vectors of which contain more than sixty-four elements must be programmed by the user as a sequence of vector instructions in a loop.

Vector load (store) instructions are provided to load (store) data from (to) contiguous or arbitrarily but equally spaced words in memory. Thus the CRAY I has no affinity to either row or column vectors. However, the sparse matrix capability of the CRAY is limited.

Vector compare instructions may be used to create a vector mask. A mask thus created or read from memory may be utilized in the vector mask register (VM) to control data movement between vector registers but cannot be used to control "gather/scatter" loads or stores from memory. The user may, however, program random loads (stores) from (to) memory and the vector registers using scalar instructions.

The CRAY I does not have specific macro instructions for inner product, sum or max/min reductions, etc.

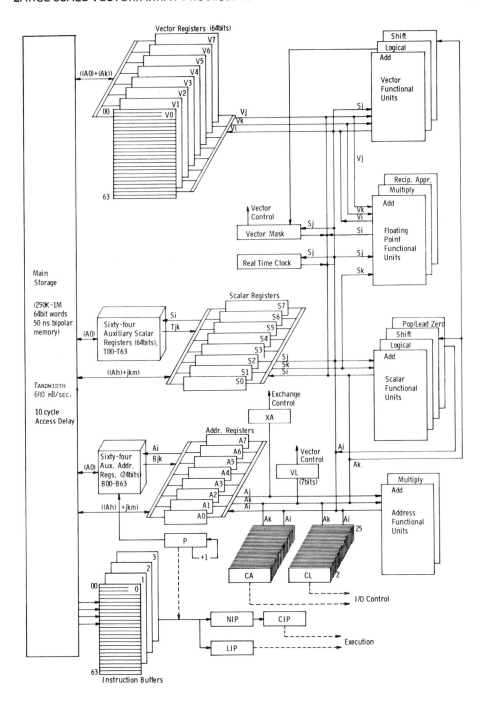

Figure VI. The Cray I Machine Organization

The machine with its limited instruction repertoire is straight forward in principle. The control of the machine is efficient and simple conceptually - instructions vector or scalar are not issued until all resources required by the instruction are available.

The CRAY I also offers a new capability to enhance performance particularly over sequences of vector instructions of short length. This capability is called "chaining". This concept is similar to the common data-bus concept of the IBM 360/91. Chaining allows the results of one vector operation to be immediately forwarded from its target register destination to participate as an operand in a second vector operation. Chains can be built up in such a manner as to allow the execution of three or more operations per cycle. A common chain observed in linear algebraic algorithms is vector load from memory, vector multiply and vector add. In this chain the three operations: load, multiply and add may take place concurrently. The primary significance of chaining is that each vector operation need not fully complete before the next is initiated. This then reduces the total drain time required by the sequence of pipelined instructions to the drain time required by the last instruction thereby reducing the overhead to be amortized over the chained sequence.

Another feature of the CRAY I is the provision of sixty-four auxiliary scalar "T" registers and of sixty-four auxiliary address "B" registers. These registers are available in one cycle to replace the contents of the scalar "S" registers or address "A" registers, respectively, and may be used to reduce the access time from main storage for subroutine calls, etc.

The main memory of the CRAY I is also one of its principal features. The main memory consists of two hundred and fifty thousand to one million 64-bit words. It is built of bipolar technology and has a memory cycle time of fifty nanoseconds. This memory design provides a very fast random access capability to support both the machine's vector and scalar high execution bandwidth.

The CRAY I is supported by a limited operating system at present but is planned to have a general multiprogramming operating system. An assembler and FORTRAN compiler are available. The FORTRAN compiler [16] also features vector syntactic extensions and automatic vectorization of scalar FORTRAN code.

System Comparisons

Tables II, III and IV below summarize the hardware characteristics of the CDC STAR-100, the TI-ASC and the CRAY I. Table II summarizes the general characteristics and logic technology of the respective central processing units, their instruction buffers and register complements. Table III summarizes the main memories and I/O channels. Table IV summarizes the floating-point execution units of the respective systems.

	CRAY I	TI ASC	CDC STAR
Clock Period	12.5 ns	80 ns	40 ns
Logic Technology	ECL-MECL	ECL	TCS
Interconnection	twisted wire	40 ohm coaxial	wire
Cooling	freon	water	freon
Word Size	64 bits	32 bits	64 bits
Address Space	24 bits	24 bits	48 bits
Virtual Storage	no	no	yes
Instruction Buffer	4	2 per pipe	1
Capacity	512 bytes	64 bytes/pipe	1024 bytes
Registers	512-64b Vector	16-32b Scalar	256-64b GP
	72-64b Scalar	16-32b Base	+ 64-128b
	72-24b Address	8-32b Index	stream buffer
	1-64b Mask	8-32b VPR	
	1-8b Length	48-32b buffer	
		per pipe	

Table II. Summary of Central Processing Units

	CRAY I	TI ASC	CDC STAR
Memory Capacity	2 to 8 mB	4 mB	4 to 8 mB
Memory Technology	bipolar - 1 k bits	bipolar - 256 bits	ferrite core
Memory Cycle	50 ns	160 ns	1280 ns
Memory Access Path	8 byte r/w	32 byte r/w	64 byte r/w
Interleave	16 way	8 way	32 way
Bandwidth	640 mB/sec	1600 mB/sec	1600 mB/sec
Error Correction	SEC	SEC	none
Error Detection	DED	DED	parity
Memory Extension	none	8 mB, 1000 ns cycle	none
Channels	12 in 12 out	variable w/ expander	4 to 12 +mem. chnl.
Bus Width	16 bits	256 or 32 bits	16 bits
Bandwidth	80 mB/sec	12 or 29.2 mB/sec	5 to 10 mB/sec 50 mB/sec on mem. chnl.

Table III. Summary of System Memories and I/O Channels

	CRAY I	TI ASC	CDC STAR
Function Units	12	1 per pipe	3
Segmented	yes	yes, except fltg. pt. mult. and divide.	yes
Chaining	yes	no	no
Floating Point Formats			
32 bit	no	yes	yes
radix		hex	binary
exponent		7b biased	8b signed
mantissa		24b + sign	24b incl. sign
64 bit	yes	yes	yes
radix	binary	hex	binary
exponent	15b biased	7b biased	16b signed
mantissa	48b + sign	56b + sign	48b incl. sign
Asymptotic Performance			
32 bit fltg. pt.			
additions		16 mflops/pipe	100 mflops
multiplications		16 mflops/pipe	100 mflops
divisions			50 mflops
64 bit fltg. pt.			
additions	80 mflops	9.25 mflops/pipe	50 mflops
multiplications	80 mflops	5.25 mflops/pipe	25 mflops
divisions	25 mflops	4.0 mflops/pipe	12.5 mflops

Table IV. Summary of Floating–Point Execution Units

SUMMARY

The vector instruction processors form a particularly interesting subclass of the array processors. Both conceptually and in practice, vector instructions serve as an excellent means of specifying the mathematical formulation as well as the data structure of a broad category of scientific problems of importance, particularly those problems based in numerical linear algebra and/or time series analysis. Vector approaches to the broader class of commercial applications are only now evolving, as new algorithms are being developed for sorting, tree search and data base applications.

The necessity that these machines handle sparse matrix problems using various mapping techniques for storage compression cannot be over emphasized. The range of techniques permitted should include minimally the use of Boolean arrays (bit maps) to specify positional significance as well as an "indirect addressing" facility using index vectors. Threaded list structures sometimes utilized in sparse matrix algorithms may be overly complex to retain the inherent efficiencies of the vector machine organizations.

The performance capability of any array processing system, vector or otherwise, executing conventional scalar code also cannot be over emphasized. It is rare even in the very large scale scientific applications such as numerical weather prediction or seismic data processing, etc., for these applications to contain more than eighty-five to ninety percent vector content. Furthermore, although many of these problems are increasing in size and computational complexity, this percentage is diminishing as the physical approximations or models underlying these computations are becoming more complex and pointwise dependent over the computational grid.

Lastly, the subject of software support and, in particular, standard language extensions have yet to be developed for array or vector processing systems. To date each vendor has defined and implemented language extensions particular to his machine organization. It is vitally important that a common language or language extension be developed which is opaque to any specific hardware embodiment or architecture. Likewise the question of automatic vectorization of existing scalar programs is of extreme importance, not only to aid conversion and migration of existing programs to the new hardware, but also to optimize the utilization of the hardware. Today automatic vectorization techniques are still somewhat primitive and in some cases vectorization impedes code optimization. Management of system resources in vector machines is more important, particularly storage and vector registers (if present) as the quantity of state data to be preserved during task switches and changes of subprogram contexts becomes larger.

REFERENCES

1.   C.V. Ramamoorthy and H.F. Li, "Pipeline Architecture," ACM Computing Surveys, Vol. 9, No. 1, (March 1977), pp. 61-102.

2.   K.J. Thurber and L.D. Wald, "Associative and Parallel Processors," ACM Computing Surveys, Vol. 7, No. 4, (December 1975), pp. 215-255.

3.   K.J. Thurber, *Large Scale Computer Architecture - Parallel and Associative Processors,* Hayden Book Company, Rochelle Park, N.J., (1976).

4.   S.S. Yau and H.S. Fung, "Associative Processor Architecture - A Survey," ACM Computing Surveys, Vol. 9, No. 1, (March 1977), pp. 3-27.

5.   T.C. Chen, "Unconventional Superspeed Computer Systems," AFIPS 1971 Spring Joint Computer Conference, AFIPS Press, Montvale, N.J., (1971), pp. 365-371.

6.   T.C. Chen, "Parallelism, Pipelining and Computer Efficiency," Computer Design, (January 1971), pp. 69-74.

7.   M.J. Flynn, "Some Computer Organizations and Their Effectiveness," IEEE Transactions on Computers, Vol. C-21, No. 9, (September 1972), pp. 948-960.

8.   G. Paul and M. Wayne Wilson, "The VECTRAN Language: An Experimental Language for Vector/Matrix Array Processing," IBM Palo Alto Scientific Center report G320-3334 (August 1975).

9.   G. Paul and M. Wayne Wilson, "An Introduction to VECTRAN and Its Use in Scientific Applications Programming," Los Alamos Workshop on Vector and Parallel Processors, (September 1978).

10. P.B. Schneck, Ed., "Proceedings of a Conference on Programming Languages and Compilers for Parallel and Vector Machines," ACM SIGPLAN Notices, Vol. 10, No. 3 (March 1975).

11. Control Data Corporation, *Control Data STAR-100 Computer - Hardware Reference Manual,* Manual No. 60256000.

12. Control Data Corporation, *STAR FORTRAN Language Version 2 Reference Manual,* Manual No. 60386200.

13. Texas Instruments, Inc., *Description of the ASC System - Parts 1 to 5,* Manual Nos. 934662 to 934666.

14. Texas Instruments, Inc., *ASC FORTRAN Reference Manual,* Manual No. 930044.

15. Cray Research, Inc., *CRAY-I Computing System - Reference Manual,* Manual No. 2240004.

16. Cray Research, Inc., *CRAY-I Computer System Preliminary CRAY FORTRAN (CFT) Reference Manual,* Manual No. 2240009.

Supplemental References

17. G.H. Barnes, et al, "The ILLIAC IV Computer," IEEE Trans. Computers, Vol. C-17, No. 8, (August 1968), pp. 746-757.

18. A.J. Evansen and J.L. Troy, "Introduction to the Architecture of a 288-element PEPE," Proc. 1973 Sagamore Conference on Parallel Processing, Springer-Verlag, New York, (1973), pp. 162-169.

19. P.M. Flanders, et al, "Efficient High Speed Computing with the Distributed Array Processor," in *High Speed Computer and Algorithm Organization,* edited by D.J. Kuck, et al, Academic Press, New York, (1977), pp. 113-128.

20. K.E. Batcher, "STARAN Parallel Processor System Hardware," Proc. AFIPS 1974 National Computer Conf., Vol. 43, AFIPS Press, Montvale, N.J., (1974), pp. 405-410.

21. L.C. Higbie, "The OMEN Computers: Associative Array Processors," IEEE COMPCON (1972), pp. 287-290.

22. International Business Machines, Inc., *Custom Equipment Description: 2938 Array Processor,* Form No. GA24-3519.

23. International Business Machines, Inc., *IBM 3838 Array Processor Functional Characteristics,* Form No. GA24-3639.

24. L.P. Schoene, "Master Data Processor," IBM FSD Technical Directions, Vol. 3, No. 2, (Autumn 1977), pp. 2-6.

25. G.R. Allen, et al, "The Design and Use of Special Purpose Processors for the Machine Processing of Remotely Sensed Data," Conference on Machine Processing of Remotely Sensed Data, Purdue University, October 16-18, 1973.

26. Anon, "Array Processing," Sperry Rand Engineering Vol. 2, No. 4, (1971), pp. 2-8.

27. Floating Point Systems, Inc., *AP-120B Array Transform Processor,* Manual No. FPS-7259.

28. Datawest Corporation, *Real Time Series of Micro-Programmable Array Transform Processors,* Product Bulletin Series B.

29. CSP, Inc., *An Introduction to the MAP Series - Models 100, 200 and 300.*

A LOW-COST IMAGE PROCESSING FACILITY EMPLOYING A NEW HARDWARE REALIZATION OF HIGH-SPEED SIGNAL PROCESSORS

A. Peled
IBM Scientific Center
Haifa
Israel

ABSTRACT

In this paper, we describe a low-cost image-processing facility, assembled at the IBM Israel Scientific Center in Haifa, for computer aided processing of data obtained from medical ultrasonic imaging instruments. The system comprises of (1) an IBM Series/1 (S/1) minicomputer that serves as the overall system resources manager and programmer interface, (2) an experimental low cost high speed signal processor - the Simple Signal Processor (SSP) utilizing new reduced computational complexity (RCC) signal processing algorithms for convolution and discrete Fourier transform (DFT), (3) a Ramtek 9351 gray scale/pseudo color display, (4) a 20 MHz analog to digital converter, and (5) high speed microprocessor based interfaces between the SSP, the Ramtek, and the S/1 allowing burst block transfers between any pair.

First, the two main operational modes of the system are described highlighting the interaction between the various system components. The first mode is the real-time data acquisition mode in which the ultrasound RF signal is digitized and stored as a basic image. The second mode involves processing of the raw image data for purposes of enhancement, filtering, analysis, classification, compaction or similar functions, and display or storage of the resulting images.

Next, the architectural implications of the reduced computational complexity algorithms (e.g., the Winograd DFT), which require about one fifth the number of multiplications of previous known algorithms, are discussed briefly and the SSP is described in some detail. It is shown that the use of RCC algorithms enables the SSP, which is only a 4000-circuit processor, to provide throughput rates compatible with our real-time requirements, e.g., it does a 1008 complex point DFT in about 20 msec.

Finally, several image-processing examples are discussed, mainly convolution and spectral analysis, with emphasis on processing time analysis to identify the system bottlenecks. Based on these several guidelines for future system architecture for such a facility are included.

[1]The author was with the IBM Israel Science Center, Haifa, Israel.

1. INTRODUCTION

In this paper, we describe a low-cost image-processing facility assembled at the IBM Israel Scientific Center in Haifa, for computer aided processing of data obtained from medical ultrasonic imaging instruments.

The growing acceptance of ultrasonic scanners as an effective modality for examining a variety of internal organs has spurred considerable research activity aimed at improving the diagnostic capabilities of such scanners. The apparently safe nonionizing and noninvasive nature of ultrasonic waves, combined with their ability to distinguish between various soft tissues, have led to the widespread use of ultrasonic scanners in most hospitals. They are used for a variety of medical examinations including the measuring of the performance of the heart and the flow of blood, identifying tumors, cysts and cancerous growth in various tissues, e.g., breast or liver, following development of the fetus in pregnant women from the earliest stages, detecting various abnormalities, and many more. These factors have led the Scientific Center to engage in a long-range research program aimed at improving the diagnostic capabilities of ultrasonic scanners through computer-aided processing of the data obtained. The basic idea is to determine to what degree the addition of computers to modern state-of-the-art medical ultrasonic scanners will improve their diagnostic capability. To this end, we are working in close cooperation with physicians from the Sheba Medical Center and have set up the image-processing facility described in this paper.

The ultrasonic waves used in medical diagnosis are sound waves ranging in frequency from 1.5 MHz to 15 MHz, with the most commonly used range being two to four MHz. Figure 1 depicts the basic principles of echo ultrasound. An ultrasound wave is generated by exciting a piezo-electric transducer. The sound wave enters the body and is partially reflected at boundaries that differ in their characteristic impedance. The reflected echoes, though severely attenuated, are detected by the same transducer and converted into an electric signal as shown in Figure 1. The velocity of ultrasound through various soft tissues of the body is between 1,400 and 1,600 m/sec (close to the velocity of sound in water). These small differences result in rather faint echoes; however, the receiver is sensitive enough to pick them up. In any imaging system of this type, the resolution is limited by the wavelength of the radiation; thus better resolution calls for higher frequency radiation. Unfortunately, the attenuation of the ultrasound wave increases proportionally to the frequency. Thus, for a given receiver sensitivity it is necessary to compromise between resolution and depth of penetration. In practice for abdominal and heart examination, a two-to-three MHz transducer is used, yielding a penetration of about 10 to 15 cm.

We conclude this brief introduction to medical ultrasound by describing how an image of an organ cross-section is formed. The principle is illustrated in Figure 2. A transducer is attached to a precision mechanical arm equipped with position resolvers. The physician slowly moves the transducer across the organ to be scanned. The transmitter emits pulses at a rate of about one KHz (i.e., every millisecond) and the receiver detects the echoes. The echo signal is used to modulate the intensity of a storage gray-scale CRT, with time proportional to depth, modulating the y-axis. Thus,

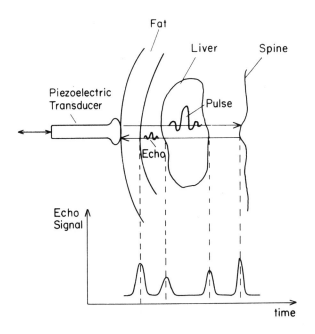

Figure 1. A Basic Echo Ultrasound Scan

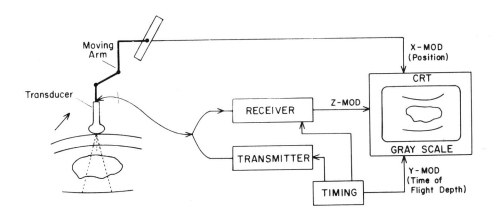

Figure 2. A Basic B-Scan System

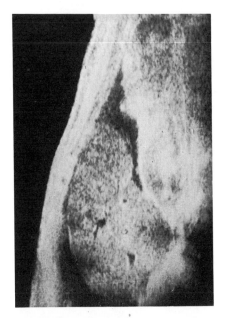

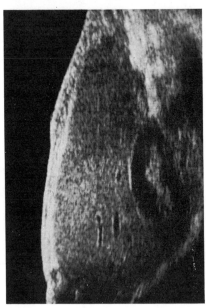

Figure 3. Typical B-scans of Liver and Kidneys,
Normal and Pathological

as the transducer is moved across, an image of the organ is registered on the display. An image obtained in such a fashion is referred to as a compound B-scan. The newest medical B-scanners are now being outfitted with a digital memory to store the image and display it, rather than an analog storage tube. However, they use only four bits per pixel leading to a rather poor utilization of the dynamic range of the received signal. A typical example of such B-scan images is shown in Figure 3.

The two-dimensional B-scan described above is not suitable for viewing moving organs, e.g., the heart. This led to the development of real-time scanners in which the transducer is rapidly moved mechanically across the organ so as to produce a series of freeze-action images displayed sequentially. This permits viewing of a cross-section of the heart in motion and observing the valve and wall motion in real time. Newer systems have replaced the mechanical motion of the transducer by electronic steering of the ultrasound beam by appropriately exciting a phased array of transducers.

Due to obvious space limitations, we have only highlighted here the features of ultrasonic imaging which are needed for following the rest of the paper. The interested reader is referred to (1, 2, 3) for a more complete discussion.

The image-processing system to be described actually has two roles. In addition to the standard mode of processing a digitized image, it also has to serve as a data acquisition system that allows digitizing and storing the returning ultrasound echoes in real time. The need for data-acquisition stems from the fact that even state-of-the-art ultrasonic scanners do not store a "good-enough" digitized version of the image to allow sophisticated post-processing algorithms to be carried out. For example, the scanner used in our system stores the image with only four bits per pixel after considerable nonlinear processing of the signal which is irreversible. To allow us to experiment with a variety of signal processing algorithms, we sample the echo signal with an eight-bit A/D converter directly after the preamplifier and before any analog processing. Considering that the data has to be sampled at rates between 2 to 20 MHz and some 250,000 pixels gathered in less than a second, a task that strains even large and expensive computers, we had to invest a considerable effort to accomplish this task without driving up the cost of the system.

In the rest of this paper, we describe the system concentrating on the signal processing aspects. In Section 2 we describe the basic components of the system and their interconnection. We then proceed to explain the two basic operational modes, data acquisition and data processing. Section 3 is devoted to highlighting the signal processing aspects of the system, in particular we describe the Simple Signal Processor (SSP), a low-cost experimental signal processor capitalizing on recently derived reduced computational complexity (RCC) signal processing algorithms. The SSP serves in the data acquisition as well as in the number crunching as the high-speed processor of the whole system.

Finally, in Section 4 several image processing examples, mainly convolution and spectral analysis, are discussed in some detail highlighting the various capabilities of the system.

2. IMAGE-PROCESSING SYSTEM ARCHITECTURE

The overall system architecture is depicted in Figure 4. At the heart of the system serving as the overall system coordinator and main user interface is the IBM Series/1 minicomputer. It is equipped with 128 K bytes of main storage with a 660 nsec cycle time. Its CPU (4955 Processor) has an average instruction time (weighted) of 3.9 μsec, four priority interrupt levels with eight general purpose registers per interrupt level. Its I/O channel accomodates data transfers in burst cycle steal at rates of 1.6 Mbytes/second. It also has an analog sensor input equipped with a 14-bit analog-to-digital converter. A 14-Mbyte, fixed-head, nonremovable disk serves as mass storage, and 0.5 Mbyte removable diskettes serve for long-term storage of various images. The availability of a "Realtime Programming System" operating system and the ability to write application programs in PL/I, a high-level language, make the S/1 a convenient choice for its function as system controller.

The S/1 communicates with, and controls the Ramtek 9351 graphic display and the SSP, a high-speed signal processor, through two microprogrammable interfaces the Channel Ramtek Adapter (CRA) and Channel SSP Adapter (CSA), respectively. These two interfaces which are basically identical, are high-speed front-end micropro-cessors handling the S/1 cycle steal channel protocol, interpreting or passing on S/1 commands or data. They have been designed using the AMD 2900[1] series of bipolar microprocessor bit slices. They have a writeable control store (WCS) which can be loaded from the S/1, thus enabling the S/1 to dynamically change their function. As we will see, these interfaces are also capable of performing local control and processing functions somewhat relieving the processing load of the S/1 and enhancing the system throughout.

The Ramtek 9351 is equipped with (512 x 512) x 16 bits of RAM storage, with a cycle time of 1.5 μsec and is microprocessor-controlled. Using the trackball the physician can outline various regions on the monitor on which further analysis is to be performed or measurements are to be taken.

The Simple Signal Processor (SSP), an experimental 16-bits fixed-point, high-speed signal processor (a 100 nsec cycle time) constructed at the IBM Thomas J. Watson Research Center, serves in a dual role. In the data acquisition mode it controls the sampling rate of the analog-to-digital converter, in our system a TRW[2] integrated TDC 1008 up to 20 MHz 8 bit A/D converter, and accumulates the samples of the RF or video signal in real time. In the interval between two successive ultrasound pulses (about 800 μsec) it dumps out the data to the Ramtek memory through the CRA. The A-reg serves as a buffer accumulating two successive eight-bit samples and forming a 16 bit word that is read in by the SSP at a maximum rate of 10 MHz, thus resulting in a maximal 20 MHz eight-bit real-time burst mode sampling

---

[1]Advanced Micro Devices Inc., 901 Thompson Place, Sunnyvale, CA 94086.
[2]TRW Electronics Systems Division, One Space Park, E2/9043, Redondo Beach, CA 90278.

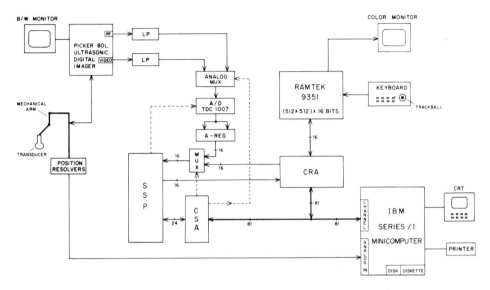

Figure 4. Image Acquisition and Processing System for Ultrasonic Data

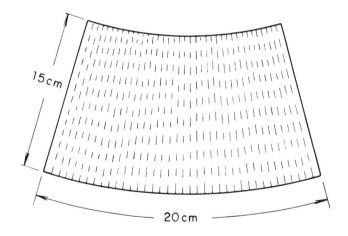

Figure 5. A Cross Sectional Scan

rate. In the data-processing mode data vectors stored either in the Ramtek image memory or in the S/1 are passed on to the SSP which performs a variety of signal processing tasks on them and returns the results. The SSP programs can be loaded by the S/1 dynamically, thus permitting a wide range of algorithms to be executed. The dotted lines on Figure 4 indicate control lines which determine the operational mode and the sampling rate.

The Picker[1] Ultrasonic Digital Imager is a standard medical scanner used routinely in hospitals. As we see, we have the option of sampling directly the RF or the video signal. The RF signal will be sampled at rates up to 20 MHz and the video signal at rates up to 2.5 MHz. The scanner has a mechanical arm equipped with precision position resolvers, their output is sampled once per pulse by the S/1 through its analog input. The Picker itself is equipped with a B/W monitor on which the compound B-scan image of the cross section of the organ scanned appears. Ultimately it will be the difference, as perceived by the physician, between the two images, the original one and the various types of computer constructed and enhanced images appearing on the Ramtek color monitor, that will determine whether computer aided processing can help improve the diagnostic capabilities of ultrasonic scanners.

Before we proceed to describe the two main operational modes of the system, we have to provide some additional details on the CSA and CRA. As mentioned above they are bipolar microprocessors capable of handling the S/1 channel protocol, that is initiating interrupts, executing direct program control (DPC) commands from the S/1, and sustaining burst-mode cycle steal data transfers at the maximal channel rate of 1.6 MByte/sec. They operate at a 132 nsec cycle time with one microinstruction executed in every cycle. Each interface can receive external interrupts and act upon them immediately (in the next cycle). A more complete description of the CSA and CRA interface is given in the Appendix and the reader is advised to consult it for details of their operation.

2.1 The Data-Acquisition Mode

In the data-acquisition mode, the system has to sample and store in real time a compound B-scan ultrasound image. A typical example that we consider is the case that ultrasound pulses are emitted every one msec and echoes are received for 200 μsec, corresponding to about 15 cm depth of penetration. A 2.9 MHz sampling rate of the video output will result in 400 words (eight-bits each) per returned echo.

To cover a cross-section as shown in Figure 5, that is about 20 cm wide at the far end, with a 0.5 mm lateral resolution, we need to store 400 scan lines to form the ultrasound image. Since the pulse repetition rate of the ultrasound scanner is one KHz and an average scan takes 1.5 to 3 seconds, we will obtain several returned echoes for each scan line that we wish to represent. For exmaple, for a lateral movement of the transducer of 10 cm/sec, we will get about five returns for each scan line finally shown in the CRT. These several returns are averaged by the SSP to produce a single

[1]Picker Corporation, 12 Clintonville Road, Northford, Conn. 06472.

scan line. Figure 6 depicts the image acquisition process showing the interaction between the various system components. Each pulse transmitted by the ultrasound scanner generates an interrupt to the CSA (every one msec), which causes the SSP to start sampling the video data at the 2 MHz rate into an input buffer. At this point it is possible to introduce a delay, by a proper CSA program that will idle a given amount of time before sending the "Start SSP" command. When the input buffer is filled in the SSP it interrupts the CSA. The CSA, in the meantime, has received a command from the S/1 based on the "Position Computation" whether to instruct the SSP to accumulate the received echo, or to send out the previous scan line through the CRA to the Ramtek. Both cases are shown in Figure 6. As we see from the timing diagram, about 750 μsec are available for outputting data to the Ramtek. If no packing is done by the SSP (i.e., two eight-bit words put in a 16-bit word), 500 words can be sent in this time interval, based on 1.5 μsec Ramtek memory cycle time. This corresponds to a maximal 2.5 MHz sampling rate of a full echo lasting 200 μsec, i.e., 15 cm depth.

A second data acquisition mode is also used for purposes of tissue characterization. In this mode only echoes from a limited local area of the cross-section (up to 4 × 4 cm) are sampled from the RF port at 10 MHz rates, or in special cases, at 20 MHz rates, by packing the data before input to the SSP in the A-Reg. In this mode each returned echo is passed on to the Ramtek memory in the 750 μsec interval between two subsequent returns. This data is not used to display an image, but to perform spatial spectral analysis for characterizing the particular place of tissue being analyzed. The area to be analyzed in this fashion is defined by the operator (physician) via the Ramtek traced mode by outlining a rectangle with the cursor on the monitor. The S/1 computes the absolute position of the transducer corresponding to this area, and the physician performs a scan during which, upon detection of the proper position (within a certain accuracy), the RF data acquisition is initiated.

2.2 The Data-Processing Mode

In this operational mode it is assumed that the data to be processed resides in the Ramtek memory, e.g., an image, and the operator wishes to perform a certain type of processing of this image. A simple example may be computing the histogram, and displaying it on the Ramtek monitor. More complicated functions include filtering of the image, or spatial spectral analysis. This mode is relatively speaking simpler than the data-acquisition mode, since there is no real-time dependency. The S/1 instructs the CRA and the SSP to execute a block transfer between the Ramtek memory and the SSP, and instructs the SSP via the CSA what function (i.e., which SSP program is to be executed for this vector. The SSP, when completing the execution, interrupts the S/1 via the CSA and confirms completion, upon which the S/1 fetches the results. By using the direct memory access through cycle steal into the SSP and using double buffering of the data in the SSP memory, the data transfer cycles and execution are overlapped improving the throughput considerably. In our system, the SSP operates on a 100 nsec cycle time, while the maximum data rate into the Ramtek memory is 1.5 μsec, i.e., a ratio of 15. This implies that as long as the computational function to be executed by the SSP requires more than 15 cycles/point, the function is compute bound and virtually no time is lost on I/O owing to the overlap used. The 4 K word

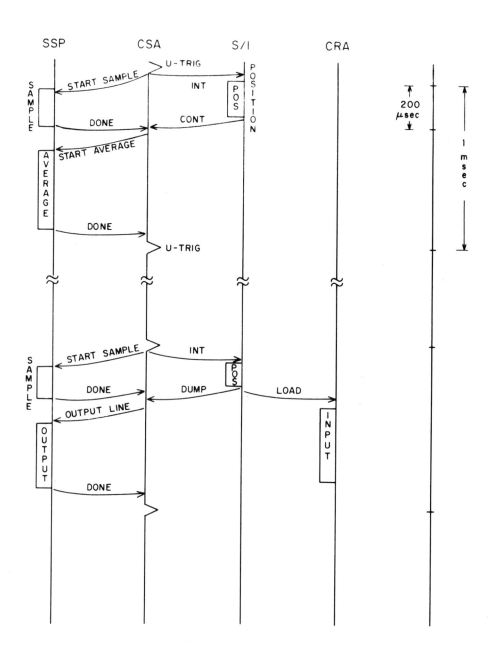

Figure 6. Image Acquisition Timing

data memory of the SSP allows such double buffering in most processing cases. Figure 7 depicts the timing of the data-processing mode, showing the overlap between the processing and the I/O.

Finally, we should mention that the same type of processing is possible when the data resides in the S/1 and the results are to be returned to either the S/1 or the Ramtek.

3. THE SIMPLE SIGNAL PROCESS (SSP)

The disclosure by Winograd (4) of a new algorithm for the computation of the DFT which requires about one-fifth the number of multiplications of the by now "standard" FFT algorithm, and approximately the same number of additions, has signalled the beginning of a new approach to the derivation of computational algorithms for digital signal processing. It involves using results from the theory of computational complexity, a relatively young field concerned with the simplification of the computational tasks required to evaluate various mathematical expressions (5). This approach has been since used by additional researchers in the field to obtain similar algorithms for convolution and DFT (6, 7, 8). The architectural implications of these new algorithms as well as those of the previous FFT algorithms are still not fully understood, which spurred the development of a whole class of special purpose processors that were optimized for the FFT algorithms. However, it became quite clear as these new Reduced Computational Complexity (RCC) were derived that they will not benefit existing signal processors that have been optimized for the FFT algorithm. This is due to the mismatch between the RCC algorithm features and the architecture of such special processors (e.g., the SPS 41/81[1], MAP-100/200[2], AP-120B[3], IBM 3838[4]). The three main factors contributing to the mismatch are:

1. While existing processors are usually equipped to do bit reversal of the data upon input to the processor as required by the FFT, the reordering required for the RCC algorithms is according to the Chinese remainder theorem and is considerably more complex.

2. While existing processors employ hardware parallelism of about two adders and one multiplier, the ratio in FFT algorithms, the add/multiply ratio in RCC algorithms, is about eight to one and they cannot be easily overlapped.

3. Existing processors employ arithmetic pipelining to increase the throughput. Its efficient utilization is made possible by the highly regular and symmetric flow of computations in the FFT algorithm. The RCC algorithms are characterized by a considerably more complex and less regular flow of computation which tends to empty the pipeline often, thereby decreasing its efficiency.

---

[1]Signal Processing Systems, Inc., 223 Crescent Street, Waltham, Mass. 02154.
[2]CSP inc., 209 Middlesex Turnpike, Burlington, Mass. 01802.
[3]Floating Point Systems, Inc., P.O. Box 3489, Portland, Or. 97223.
[4]International Business Machines Corp., Armonk, N.Y. 10504.

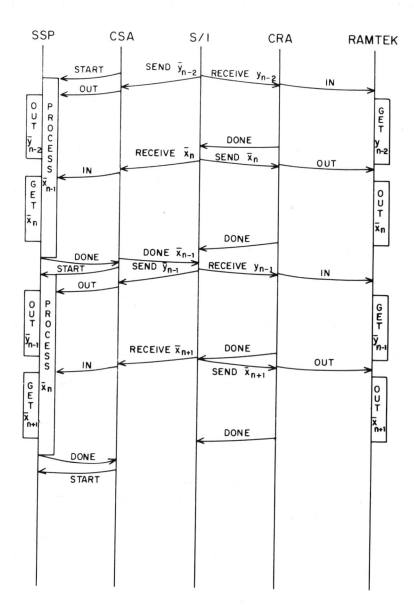

Figure 7. Data-Processing Timing
(Vector $\bar{x}_n$ is processed to result in $\bar{y}_n$)

The net effect of the mismatches outlined above is that the RCC algorithms do not benefit existing specialized computers for signal processing since their use does not lead to their cost/performance improvement, quite to the contrary, it will even degrade their throughput.

In light of these factors, the experimental Simple Signal Processor (SSP) was designed at the IBM Thomas J. Watson Research Center as a low cost signal processor for general purpose signal processing applications. Its performance is significantly improved by the utilization of RCC algorithms to implement the basic signal processing kernels of DFT and convolution. Furthermore, the term "simple" refers not only to the hardware, but to the software aspects as well. While existing special signal processors have to be programmed using horizontal time dependent microcode, the SSP uses a simple one operand instruction format designed so as to permit easy compilation from a high level language without loss of performance.

Figure 8 depicts the block diagram of the SSP. Basically, the SSP is a general purpose computer intended for signal processing applications operating under a host computer control. However, it is important to note that the SSP can carry out a complete signal processing application without any host intervention and is different in this respect from the Arithmetic Processors employed in current signal processors. The arithmetic unit (AU) design does not incorporate a hardware multiplier and performs multiplications of data by fixed coefficients using the canonical signed digit code (CSDC) and multiplications of data by data using Booth's algorithm with two bits at a time. The merits of such a design are discussed in detail in (9). The SSP has in addition to the AU three other hardware entities which participate in the execution of instruction. They are: the instruction store and associated instruction sequencing mechanism, the operation decode and effective address calculation unit, and the data RAM. The SSP instruction execution is pipelined in four phases in which each of the entities listed above is active in execution of subsequent instructions. This is depicted in Figure 9. The SSP has 64 machine instructions, most of which take exactly four machine cycles to complete yielding a throughput of $1/T_c$, where T_c is the SSP cycle time. The only exceptions are the MPY and DIV (multiply and divide) instructions that take eight and 16 cycles, respectively for a 16-bit machine. Obviously, some instructions, e.g. conditional branch, may require more cycles due to the fact that the condition may still be computed in the pipe. However, since most signal processing computations are not data dependent, but consist of the execution of a fixed computation sequence, the occurrence frequency of conditional branches of the type that stall the pipeline is rather low, resulting in only a minor degradation in throughput. It is important to note that conditional branches on the index register, of the type used in closing a loop, do not stall the pipeline.

The instruction format is simple, having a 6-bit OP field which defines one of 64 instructions and a 14-bit address field which gives an operand address. Most instructions execute on an operand fetched from the effective address calculated using the index register and a compute pointer. The mask features of the index register described in (10) which are incorporated in the SSP allow efficient manipulation of circular buffers in memory.

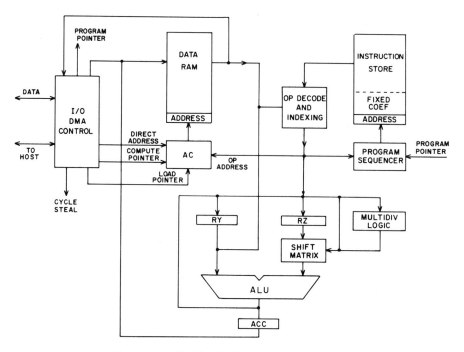

Figure 8.   The Simple Signal Processor (SSP)
Block Diagram

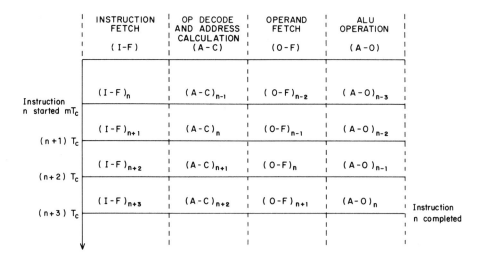

INSTRUCTION FETCH (I-F)	OP DECODE AND ADDRESS CALCULATION (A-C)	OPERAND FETCH (O-F)	ALU OPERATION (A-O)	
$(I-F)_n$	$(A-C)_{n-1}$	$(O-F)_{n-2}$	$(A-O)_{n-3}$	
$(I-F)_{n+1}$	$(A-C)_n$	$(O-F)_{n-1}$	$(A-O)_{n-2}$	
$(I-F)_{n+2}$	$(A-C)_{n+1}$	$(O-F)_n$	$(A-O)_{n-1}$	
$(I-F)_{n+3}$	$(A-C)_{n+2}$	$(O-F)_{n+1}$	$(A-O)_n$	Instruction n completed

Instruction n started mT_c

$(n+1) T_c$

$(n+2) T_c$

$(n+3) T_c$

Figure 9. The Pipelining of Instruction Execution in SSP

One of the imporant features of the SSP is the way in which multiplications by fixed coefficients are performed and therefore it will be described in more detail. As was already mentioned, this is done using CSDC representation for the coefficients (9). This is really an extension of the notion of "preconditioning" employed in the derivation of RCC algorithms, i.e., taking advantage of the fact that the coefficients are known in advance of run time and can be translated at compile time into a sequence of shift/add instructions that will insure that multiplication by these coefficients is executed in the minimum number of machine cycles. Table I depicts the sequence of machine instructions generated to calculate:

$$p_i = a \cdot x_i + b \cdot y_i + z \cdot c_i \tag{1}$$

	99	$RX \leftarrow 0$
	100	$RZ \leftarrow x_i$
	101	GOTO COEFFICIENT SUBROUTINE; PUSH (PC+1)
	102	$RZ \leftarrow y_i$; PUSH (PC+1); PC $\leftarrow$ SAVE
	103	$RZ \leftarrow z_i$; PUSH (PC+1); PC $\leftarrow$ SAVE
	104	$RX \leftarrow RX+1$
	105	$P_i \leftarrow RY$
	106	$RX < 100$
	107	YES LOOP BACK GOTO 100.
	200	$RY \leftarrow RZ \cdot 2^{-2}$
	201	$RY \leftarrow RY + RZ \cdot 2^{-5}$
xa	202	$RY \leftarrow RY - RZ \cdot 2^{-9}$
	203	$RY \leftarrow RY + RZ \cdot 2^{-13}$; POP PC; SAVE$\leftarrow$PC+1;
	204	$RY \leftarrow RY + RZ \cdot 2^{-3}$
xb	205	$RY \leftarrow RY - RZ \cdot 2/^{-7}$; POP PC; SAVE$\leftarrow$PC+1;
	206	$RY \leftarrow RY + RZ \cdot 1^{-1}$
	207	$RY \leftarrow RY + RZ \cdot 1^{-5}$
xc	208	$RY \leftarrow RY + RZ \cdot 2^{-7}$
	209	$RY \leftarrow RY + RZ \cdot 2^{-10}$
	210	$RY \leftarrow RY + RZ \cdot 2^{-12}$; POP PC; SAVE$\leftarrow$PC+1;

$a = (0.0\ 1\ 0\ 0\ \overline{1}\ 0\ 0\ \underline{0}\ 0\ 1\ 0\ 0\ 0\ 1)_2 = (0.236450195)_{10}$
$b = (0.0\ 0\ 1\ 0\ 0\ 0\ \underline{1})_2 = (0.117189)_{10}$
$c = (0.1\ 0\ 0\ 0\ 1\ 0\ 1\ 0\ 0\ 0\ 1\ 0\ 1)_2 = (0.5227050781)_{10}$

Table I. The SSP Program Required to Execute a
$P_i = (a \cdot x_i + b \cdot y_i + c \cdot z_i)$ Sum of Products.

which is a typical sum of products that is one of the main kernels being executed in signal processing. The overhead required in going to a list of coefficients is only one instruction and the subroutine linkage machine is such that no further machine cycles are wasted, and the total number of machine cycles required to execute the sum of products is directly proportional to the number of nonzero digits of the fixed coefficients (a_N). In (9) it was shown that on the average $a_N \simeq B/3$, where B is the number of bits used to represent the coefficient, and for FIR filter taps the $a_N \simeq B/5$. Thus, to implement such a sum of N products on SSP will require P_c machine cycles per output, where P_c is given by

$$P_c = N(a_N + 1) + 3 \tag{2}$$

with the average number being (the 3 in Eq. (2) is due to the overhead in causing the loop)

$$\overline{P} = N(B/5 + 1) + 3. \tag{3}$$

Equation (3) also points out an important property of the SSP architecture the the ability to directly trade coefficient precision for running time. Decreasing the precision B of fixed coefficients will reduce the running time.

The SSP architectural features which are relevant to the remainder of the paper are summarized briefly below:

1. The SSP is a simple low-cost, 4000-circuits processor employing no arithmetic parallelism and pipelining; it does use an instruction pipeline to improve hardware utilization and throughput.

2. The time required to carry out a given computation is directly proportional to $(N_a + a_N \cdot N_{FM} + 8 \cdot N_{DM})$, where N_a is the number of additions, N_{FM} the number of multiplications by fixed coefficients having an average of a_N nonzero digits, and N_{DM} is the number of data by data multiplications. This time is nearly independent of the structure of the algorithm used.

3. It employs a simple I/O mechanism for data in direct memory access cycle steal. The rate of the transfer is determined by the SSP program as a multiple of the cycle time. Double buffering for computation and I/O is facilitated by the use of separate pointers settable by the host computer.

To illustrate the savings made possible by the RCC algorithm, we consider the implementation of a 256 point FFT versus a 240 Winograd DFT (WFT), on SSP. To evaluate a 256 point FFT on complex data 2304 M (multiplications) and 5248 A (additions) are required, whereas a 240 point WFT requires 648 M and 5016 A. When coding these algorithms on the SSP, it was found that the FFT requires about 63,400 machine cycles, while the WFT requires only 31,900 machine cycles, i.e., an almost two to one speed improvement. Furthermore, the WFT also requires only half as much overall storage (program + data) than the FFT algorithm. This is mainly due to the lower number of coefficients required in the WFT. Since the high-speed storage cost is a significant portion of the total processor cost, this is an additional important advantage. Similar improvements are obtained for a 1008 point WFT versus a 1024 point FFT, and other factors, and for convolution algorithms as well.

At this point, it is worthwhile mentioning that the availability of "cheap" one I.C. high-speed multipliers (e.g., the 16 × 16-bit multiplier announced by TRW) that would accomplish the multiplication in two machine cycles (200 nsec), does not alter the validity of the architectural decision. This is illustrated by the fact that incorporating such a multiplier, which the architecture provides for as long as addition and multiplication do not occur in parallel, will result in an only 10% additional speed improvement in the WFT. This is obviously due to the low multiply/add ration in the WFT algorithms. On the other hand, the addition of such a multiplier will almost double the SSP circuit count, which is the governing factor for an LSI implementation.

4. SOME PROCESSING EXAMPLES

To illustrate the potential capabilities of the system described, we discuss in some detail several examples of typical image-processing tasks.

4.1 Histogram Equalization

The first and basic step in many image-processing applications is to compute the histogram of a given image. Let the matrix $I(j, k)$ represent the ultrasound image, where $0 \leq 1 \ (j, k) \leq 255$, due to the eight-bit quantization of the data, and the size of the image is, say, 400 × 400, i.e., 160,000 pixels. The histogram is given by the vector $h(j)$ in which the value of the j-th element represents the number of times the pixel of size j appears in the image, and is defined by

$$h(j) = \sum_{m=0}^{N-1} \sum_{k=0}^{N-1} C_j \ [I(m,k)], \tag{4}$$

where $C_j(m)$ is defined as ($N = 400$ in our example)

$$C_j(m) = \begin{cases} 1 & j = m \\ 0 & \text{otherwise} \end{cases}. \tag{5}$$

Usually, one deals with the normalized histogram $p_h(j) = h(j)/N^2$ which is an approximation to the probability density function of the image. To compute the histogram of an image resident in the Ramtek memory the operator initiates a S/1 program that sends the image one row at a time (400 words) from the Ramtek to the SSP through the CRA. In the SSP the vector, $p_h(j)$ is being computed, and after all the rows have been sent the vector $p_h(j)$ is sent to the Ramtek for display. The SSP program to compute the histogram requires 13 cycles per pixel. This is achieved by the SSP ability to use the value of the pixel element to address the element of $p_h(j)$ to be incremented in a rather tight loop. It is possible to shorten this to seven cycles per pixel, at the expense of additional program memory by unrolling the tight loop if speed is important; however, we immediately note that computing the histogram is a task that is I/O bound in our system, and therefore speeding the computation will be only an exercise in futility. The reason for this is that, as we recall from Section 2, the Ramtek memory cycle time is 1.5 μsec which corresponds to 15 SSP cycles; thus the ability to finish our computation on a row of 400 elements is less than 600 cycles (i.e., 0.6 msec) will only mean that the SSP will wait for I/O completion before proceeding to process the next row.

A histogram of the full image will therefore be computed in less than 250 msec, allowing for system overhead. We believe this time to be sufficiently fast for good human interactive processing. It should be noted that it is at least an order of magnitude faster than the execution of such functions on similar commercially available systems.

Typically the next step after the histogram has been computed involves an attempt to equalize the distribution of the various amplitude levels, thus providing an enhanced contrast image. This procedure, usually referred to as histogram equalization, involves a nonlinear mapping of the pixel amplitude levels. As an example, we consider here the histogram hyperbolization technique (11), which is a nonlinear transformation intended to produce an equalized distribution of the pixel amplitude levels as perceived by the human eye. The transformation to be performed on each pixel $I(k,l)$ is given by

$$J\ [I(k,l)] = c\ (\exp\ [\log\ (1 + \frac{1}{c})\ \sum_{j=0}^{I-1} p_h\ (j)] - 1),\qquad (6)$$

where c is a constant and $p_h(j)$ is the PDF function computed before. To accomplish this transformation, we again have to pass the full image through the SSP, say one row at a time, and compute the transformation given in Equation 6, thus obtaining an equalized image $J(k,l)$ to be returned to the Ramtek for display. As we recall the I/O to send and receive each row to the SSP requires at least 12 msec (800 × 1.5 μsec), thus for a perfect balance, i.e. computation time equals I/O time. The transformation of 400 points should take 12,000 SSP cycles, or 30 cycles per point on the average. To accomplish this the SSP program first computes the cumulative distribution function $F_a(I)$ as

$$F_h\ (I)\ =\ \sum_{j=0}^{I-1} p_h\ (j)\qquad (7)$$

and stores it instead of the PDF function $p_h(j)$. This requires 132,684 SSP cycles, or roughly 14 msec, and is done once per image. Next, as each row is passed into the SSP, the transformation is done. The exp function is implemented using a look-up table facility incorporated in the SSP and therefore only 20 cycles per point are needed making this again an I/O bound task. The overall histogram equalization will be accomplished in about 0.5 sec including set-up overhead. This is again a sufficiently good response time for human interaction.

4.2 Spectral Analysis and Filtering

In many image-processing applications, it is required to compute the two-dimensional discrete Fourier transform (2D DFT) of an image, either for purposes of spectral analysis, or as an intermediate step in image restoration through filtering. The 2D DFT of an image is defined as:

$$F_I(k,p) = \sum_{m=0}^{N-1} \sum_{n=0}^{N-1} I(m,n) \ \exp\left(-j\frac{2\pi}{N}(km + pn)\right) \quad 0 \leq k,p,m,n \leq N - 1. \quad (8)$$

It is well known that it can be computed by repetitive execution of one dimensional DFTs as follows: We first compute the 1D DFT on each column of $I(m,n)$ (read data)

$$A(m,p) = \sum_{n=0}^{N-1} I(m,n) \ \exp\left(-j\frac{2\pi}{N}pn\right) \quad (9)$$

and then compute the 1D DFT on each resulting vector $A(m,p)$ (complex data):

$$F_I(k,p) = \sum_{m=0}^{N-1} A(m,p) \ \exp\left(-j\frac{2\pi}{N}km\right). \quad (10)$$

Considering the images obtained in our system, which are 400 × 400 the 2D DFT is computed using RCC DFT algorithms for 420 point vectors. The time required is 3.2 msec for real data and 5.8 msec for complex data. To stay within the storage limitations the complex data is kept with eight-bits for the imaginary and eight-bits for the real part and is packed into one 16-bit word before being sent to the Ramtek for intermediate storage. We note that the 1D DFT is a compute bound task, as it takes only 1.3 msec to pass the data into the SSP and back out to the Ramtek, and it takes either 3.2 or 5.8 msec to perform the computation. Based on the above completing a 2D DFT on a 420 × 420 image will take about four sec including all the set-up times. This time is again at least an order of magnitude faster than possible in similar low cost systems and short enough to insure good human interaction.

In the case that our aim is to determine only the spectral properties of the image under consideration, the processing is almost completed once the 2D DFT has been computed. The only additional step is to compute the magnitude of $F_I(k,p)$ and display (and in some cases the phase too), which is straightforward and does not add any significant delay. However, in many applications the 2D DFT computation is only one step in an image restoration process which inlcudes:

1. Compute the 2D DFT of the image;

2. compute the 2D DFT of the point spread function (PSF) matrix,

(i.e., the impulse response of the corrective filter);

3. divide the resulting 2D DFTs from steps 1 and 2; and

4. compute the inverse 2D DFT to obtain the filtered image.

Such a processing chain will require about 13 sec which is sufficiently fast to allow the user to interactively try various types of filtering to obtain the best subjective result.

5. SUMMARY

In this paper we have described an experimental low cost image acquisition and processing facility for medical ultrasound. The system serves a dual role: real-time speed image data acquisition and processing the digitized images at speeds allowing good human interactive processing. The key to achieving these goals within the limitations of a low-cost system is the SSP, an experimental high-speed processor utilizing RCC algorithms. Its major features and design philosophy were outlined in Section 3. The performance level achievable by such a system is illustrated via several processing examples in Section 4.

ACKNOWLEDGEMENT

The author wishes to acknowledge the contribution of Dr. Israel Berger and Mr. Bernard Mezrich to the overall system design that has emerged and was consolidated through our numerous stimulating discussions on the subject. Many thanks are also due to Professor Josef Raviv who was instrumental in initiating this research project and is lending it his continuing support.

REFERENCES

1.   G.B. Devey and P.N.T. Wells, "Utrasound in Medical Diagnosis", Scientific American, May 1978, pp. 98-112.

2.   Barry B. Goldberg, "Diagnostic Uses of Ultrasound", Grune & Stratton, New York, 1975.

3.   K.R. Erikson, F.J. Fry and J. P. Jones, "Ultrasound in Medicine - A Review", IEEE Trans. on Sonics and Ultrasonics, Vol. SU-21, No. 3, July 1974.

4.   S. Winograd, "On Computing the Discrete Fourier Transform", Proc. Math. Acad. Sci., U.S. April 1976.

5.   "Complexity of Sequential and Parallel Numerical Algorithm", J.F. Traub (editor), Academic Press, New York, 1973.

6.   R.C.Agarwal and J.C. Cooley, "New Algorithms for Digital Convolution",
       Proc. 1977 Intl. Conf. Acoust. Speech and Signal Processing, Hartford, Conn.,
       IEEE pub. 77CH1197-3 ASSP.

7.   D.P. Kolba and T.W. Parks, "A Prime Factor FFT Algorithm Using High-
       Speed Convolution", IEEE Trans. on Acoustics, Speech and Signal Processing,
       Vol. ASSP-25, No. 4, August 1977.

8.   H.J. Nussbaum and P. Quandalle, "Computation of Convolution and Discrete
       Fourier Transforms by Polynomial Transforms", IBM Journal R&D, Vol. 22,
       No. 2, March 1976.

9.   A. Peled, "On the Hardware Implementation of Digital Signal Processors",
       IEEE Trans. Acoustics, Speech and Signal Processing, Vol. ASSP-24, No. 1.,
       February 1976.

10.   A. Peled, "A Digital Processing Subsystem", 1976 International Conference on
       Acoustics, Speech and Signal Processing, April 1976, Philadelphia, IEEE Publ.
       76CH1067 -ASSP.

11.   Werner Frei, "Image Enhancement by Histogram Hyperbolization," Image
       Processing and Computer Graphics, Academic Press, October 1977.

APPENDIX

In this Appendix we describe briefly the general CSA and CRA architecture.
Basically, they are microprocessor based microprogrammable interfaces to the S/1
channel, whose primary function is to permit data transfers to/from the S/1 memory
in burst cycle steal mode at the maximum possible rate of 660 nsec/byte. The
decision to choose a microprocessor implementation rather than hardwired random
logic was made to allow maximum flexibility in system use, and insure future ability to
interface to other device types. Furthermore, the use of microprocessors also allows
performing local operations and decisions, thus relieving the S/1 realtime workload.
The two adapters are basically identical and differ only in the details of the device
interface registers. The operational speed required of the interface, i.e., accomplishing
several things in less than 660 nsec, dictates the use of bipolar microprocessor bit
slices.

Figure A.1 depicts the interface architecture. As evident from it we used the
AMD 2900 microprocessor bit slice series, in particular, the AM 2901 CPE - a 4-bit
central processing element with a two address architecture, and 16 general registers file
combined with an eight function ALU, and the AM 2910 sequencer - a 12-bit micro-
program sequencer capable of selecting the next address from the program counter, an
internal register, a five level deep stack for subroutine linkage, or a direct input. These
are the main LSI components of the system, it is augmented with standard high-speed
bipolar RAM (256 × 8) and PROM (256 × 8), and additional MSI components as
registers, buffers, latches, etc. Not shown in the diagram of Figure A.1 is the path

that permits loading of the μprogram RAM from the S/1 interface under the control of the program resident in the bootstrap PROM.

A μprogram sequence is initiated by the appearance of an external interrupt. For example, when the S/1 wishes to start communications with the interface, it raises one of its lines. This is recognized by the JMP CTL (jump control circuit) in the interface which causes the sequencer to branch to the address indicated by the IP PROM (interrupt PROM) which contains the absolute address of the appropriate interrupt handler routine. In our example, the interface will read at first the address bus coming from the S/1 to determine the device address and the command to be passed to it, this may be the interface itself or the device it controls. Based on the command address, the interface will now read (or write) the data from/to the S/1 data bus and pass it to or from the device addressed.

The microprocessor operation is controlled by the microinstruction fetched and latched in μIR. A new microinstruction is fetched every cycle, i.e., every 132 nsec. It has several fields controlling concurrently various portions of the processor. The main fields are:

1. Control - This field determines which external registers are read/written by the CPE.

2. I-field - This field contains an immediate operand to be passed to the CPE.

3. μCTL - This field contains the two addresses defining the two registers to be used by a CPE operation (2×4 bits) and a 9-bit code determing the ALU function and its sources and destination.

4. B-OP - This field contains one of eight possible branch instructions, including; conditional ones on ALU results from the previous cycle, nonconditional and subroutine jumps or returns.

5. B-T - This field contains the eight-bit address for a branch operation specified. Typically, the operation of the interface consists of the following sequence:

    a. The S/1 initiates a command alerting the interface to an upcoming event and passes it the appropriate parameters to be used in conjunction with this event.

    b. The occurrence of the external event is indicated by an interrupt to the interface (e.g., a signal from the SSP denoting completion of a computational task). The interrupt handler uses the parameters supplied previously by the S/1 to respond to the device (e.g., it initializes the SSP for the next task), and also notifies the S/1 (through an interrupt) that the expected event occurred and was successfully completed.

    c. The interface returns to the wait state.

The sequence outlined above points out the main strength of such a programmable interface; it allows the decoupling of the S/1 response time to the attached
device from the response time as seen by the device. That is to say, the S/1 prepares
its responses to an event in the interface over a time period, say several tens of
microseconds, and when the event occurs, the interface responds to it for the S/1 in,
say, a tenth of that time.

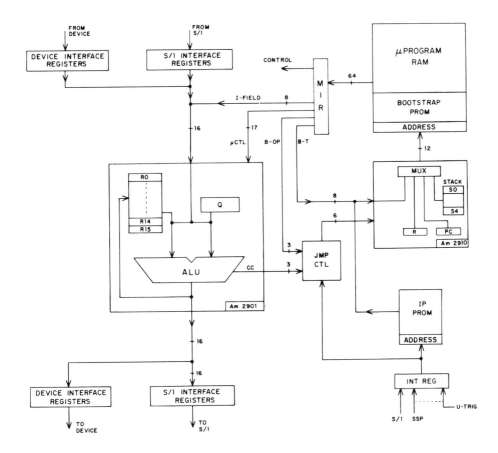

Figure A.1. Microprogrammable Microprocess Based
S/1 Interface Block Diagram

SUBJECT INDEX